A Dictionary of Named Effects and Laws

A Dictionary of Named Effects and Laws

in Chemistry, Physics and Mathematics

D. W. G. BALLENTYNE

B.Sc., Ph.D., F.Inst.P.

and

D. R. LOVETT

M.Sc., A.R.C.S., D.I.C., Ph.D., M.Inst.P.

FOURTH EDITION

LONDON NEW YORK
CHAPMAN AND HALL

First published 1958
by Chapman and Hall Ltd
11 New Fetter Lane, London EC4P 4EE
Second edition 1961
Third edition 1970
First issued as a Science Paperback 1972
Fourth edition 1980
Reprinted 1984

Published in the USA by
Chapman and Hall
in association with Methuen Inc.
733 Third Avenue, New York, NY 10017

Printed in Great Britain
at the University Press, Cambridge
ISBN 0 412 22390 2 (Science Paperback)

British Library Cataloguing in Publication Data

Ballentyne, Denis William George
A dictionary of named effects and laws. – 4th ed.
1. Science – Dictionaries
I. Title II. Lovett, D R
500.2 Q123 79-41716

ISBN 0-412-22380-5
ISBN 0-412-22390-2 Pbk

Preface to the fourth edition

The format of this edition remains unchanged from previous editions but the majority of entries have received some revision. In particular, units are now in SI units wherever possible, although with certain of the classical entries this is not possible. Chemical terminology has proved a particular problem. We have kept the common names for organic compounds because of the wide readership of this book but we have added an extra table giving the equivalent systematic names and the formulae.

We have tried to avoid omission of any *named* effects and laws that have wide usage. Nevertheless, in order to keep the book to a manageable length, it has been necessary to make a selection among the less commonly used terms and it is inevitable that some arbitrary choices and omissions must be made. Some entries from earlier editions have been left out to make room for other entries which we feel have become more important. We are especially grateful to those readers who have pointed out previous omissions.

D.W.G.B.
Imperial College, University of London
D.R.L.
University of Essex, Colchester

Preface to the first edition

Every science has its own vocabulary. It is impossible to read many pages of any scientific book without encountering words which possess a specific and unique meaning to the particular scientific subject with which the book deals. Some of these words are proper nouns, either used substantively or, more rarely, adjectivally. These are the names of scientists who have investigated a particular phenomenon or who have discovered some scientific law or relation or who have worked in some field with which their name has become historically connected.

It is with such names that this book is concerned. It is by no means intended to be read as a text-book but rather to be consulted as a dictionary whenever the reader, possibly an expert in one branch of science, is confronted by a mention of a relation or rule or law of someone or other who worked, maybe, in quite another field. He may not feel inclined to delve very deeply into the origins of the phrase. He may, in fact, wish to obtain such information as may enable him to proceed, as quickly as possible, with his reading. It is partly in an endeavour to help him that this glossary has been compiled.

Classification by subject matter has not been attempted and entries appear in alphabetical order.

Contents

Symbols

(*Unless otherwise stated, further symbols are defined within the entries.*)

c	velocity of light
e	electron charge
F	Faraday constant
g	acceleration due to gravity
h	Planck's constant
$\hbar =$	$h/2\pi$
k_B	Boltzmann's constant
m	electron mass
N	number of molecules
p	pressure
R	gas constant (per mole)
t	time
T	temperature (absolute scale)
V	volume
ε_0	permittivity of free space
μ_0	permeability of free space
exp	exponential
j =	$\sqrt{-1}$
ln	logarithm to base e
log	logarithm to base 10
O . . .	term of order . . .

Abbe Number
This is a measure of reciprocal dispersion for optical materials and is defined by

$$\nu = \frac{n_D - 1}{n_C - n_F}$$

where n is refractive index and subscripts D, C and F refer to **Fraunhofer Lines** (with wavelengths 589·3 nm, 656·3 nm and 486·1 nm respectively).

Abbe's Sine Condition
When the pencil of rays forming an optical image is of finite aperture, the condition that a magnification of the image can be obtained which is independent of the zone of the lens traversed is given by

$$\frac{n \sin \alpha}{n' \sin \alpha'} = \text{constant} = m$$

where n and n' are the refractive indices of object and image space, α and α' are the angles with which a ray leaves and (after refraction) reaches the axis again respectively and m is the lateral magnification.

Abbe's Theory (of the Diffraction of Microscopic Vision)
For the production of a truthful image of an illuminated structure by a lens it is necessary that the aperture be wide enough to transmit the whole of the diffraction pattern produced by the structure. If only a portion of the diffraction pattern is transmitted, the image will correspond to an object whose diffraction pattern is identical with the portion passed by the lens. If the structure is so fine or the lens aperture so narrow that none of the diffraction pattern is transmitted, the structure will be invisible regardless of the magnification.

Abegg's Rule
The sum of the maximum positive valency exhibited by an element

and its maximum negative valency equals 8. This rule is true generally for the elements of the 4th, 5th, 6th and 7th groups of the Periodic Table.

Abelian Group

If a group of elements A, B, C, . . . finite or infinite in number has the property that AB = BA (commutative property) for every element, then the group is known as an Abelian Group.

Abel's Identities

(1) If $A_s = a_n + a_{n+1} + \dots a_s$ then

$$\sum_{s=p}^{m} a_s b_s = \sum_{s=p}^{m} (b_s - b_{s+1}) A_s - b_p A_{p-1} + b_{m+1} A_m$$

(2) If $A'_s = a_s + a_{s+1} + a_{s+2} + \dots$ then

$$\sum_{s=p}^{m} a_s b_s = \sum_{s-p}^{m-1} (b_{s+1} - b_s) A'_{s+1} + b_p A'_p - b_m A'_{m+1}.$$

Abel's Inequality

If $u_n \to 0$ monotonically for integer values of n, then $\left|\sum_{n=1}^{m} a_n u_n\right| \leqslant A u_1$ where A is the greatest of the sums $|a_1|, |a_1 + a_2|, |a_1 + a_2 + a_3|, \dots$ $|a_1 + a_2 + \dots a_m|$.

Abel's Integral Equation

A particular type of **Volterra Equation** of the form

$$\mathrm{f}(t) = \int_0^t (t-s)^{-\alpha} \phi(s)\,\mathrm{d}s \qquad (0 < \alpha < 1).$$

The solution of the equation is

$$\phi(t) = \frac{\sin \pi\alpha}{\alpha} \frac{\mathrm{d}}{\mathrm{d}t} \int_0^t (t-s)^{\alpha-1} \mathrm{f}(s)\,\mathrm{d}s.$$

The equation, with $\alpha = 0{\cdot}5$, has particular application to the time of fall of a particle along a smooth curve in the vertical plane.

Abel's Test (for Convergence)
If Σu_n converges and $a_1, a_2, a_3, \ldots$ is a decreasing sequence of positive terms, then $\Sigma a_n u_n$ is convergent.

Abel's Test (for Infinite Integrals)
If $\int_a^\infty \mathrm{f}(x)\mathrm{d}x$ converges, and if for every value of y such that $b \leqslant y \leqslant c$ the function $\mathrm{g}(x, y)$ is neither negative nor increasing with x, then $\int_0^\infty \mathrm{f}(x)\mathrm{g}(x, y)\mathrm{d}x$ is uniformly convergent with respect to y in the range $b \leqslant y \leqslant c$.

Abel's Theorem (on Multiplication of Series)
If $c_n = a_0 b_n + a_1 b_{n-1} + \ldots + a_n b_0$, then the convergence of $\sum_{n=0}^{\infty} a_n$, $\sum_{n=0}^{\infty} b_n$ and $\sum_{n=0}^{\infty} c_n$ is a sufficient condition that $\left(\sum_{n=0}^{\infty} a_n\right)\left(\sum_{n=0}^{\infty} b_n\right) = \sum_{n=0}^{\infty} c_n$.

Abney Law
If the colour of a spectral line is desaturated by the addition of white light, its colour to the eye shifts towards the red if its wavelength is less than 570 nm and to the blue if its wavelength is greater than this value.

Adams–Bashforth Process
A method of numerically integrating ordinary differential equations. Starting with Gregory's backwards formula (*see* **Gregory's Interpolation Formulae**), $\mathrm{f}(x)$ is expanded and integrated to give

$$\frac{1}{h}\int_{x_0}^{x_0+h} \mathrm{f}(x)\mathrm{d}x = \mathrm{f}(x_0) + \left(\frac{1}{2}\nabla + \frac{5}{12}\nabla^2 + \frac{3}{8}\nabla^3 + \frac{251}{720}\nabla^4 + \frac{95}{288}\nabla^5 + \ldots\right)\mathrm{f}(x)$$

Hence, if $\mathrm{d}y/\mathrm{d}x = \mathrm{f}(x, y)$ and y and f are known for values of x up to $x = x_0$, a value for y corresponding to $x = x_0 + h$ can be obtained from f at x_0 and the backwards differences.

Agnesi, Witch of

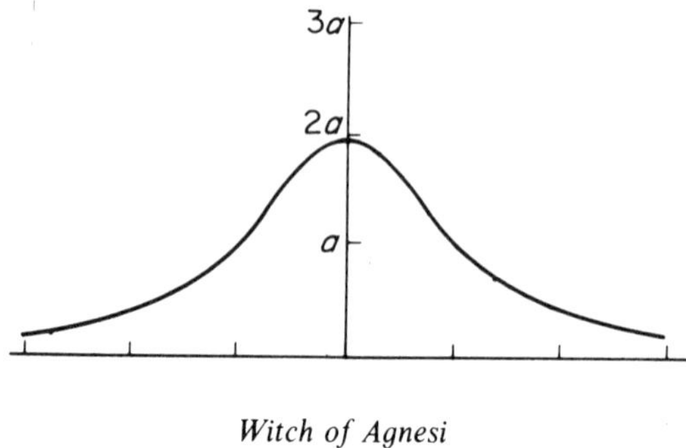

Witch of Agnesi

A curve whose equation $x^2y = 4a^2(2a - y)$.

Airy's Disc

The diffraction pattern formed by plane light waves from a point source passing through a circular aperture consists of a bright central disc, known as Airy's disc, surrounded by further rings. Airy in 1934 obtained the intensity distribution across the pattern in terms of **Bessel Functions** of order unity. The radius of the central disc is given by

$$\frac{0{\cdot}61\,\lambda}{n \sin U}$$

where λ is the wavelength *in vacuo*, U is the semi-angle of the emergent cone of light from the aperture and n is the refractive index on the image side of the aperture. The form of the diffraction pattern has particular application to the calculation of the resolving power of telescopes and other optical instruments.

Airy's Equation

An equation for multiple-beam interference for light transmitted through a plane parallel plate. The intensity of the transmitted light is given by

$$I = \frac{I_0 T^2}{(1-R)^2\left[1 + \dfrac{4R\sin^2\delta/2}{(1-R)^2}\right]}$$

where I_0 is the intensity of the incident beam, R is the reflectivity of the surfaces of the plate and T their transmissivity, and δ is the phase difference between the directly transmitted light and light which is once reflected from the two internal surfaces of the plate. (*See also* **Fabry–Pérot Fringes.**)

Airy's Integral
This is defined as

$$\mathrm{Ai}(z) = \frac{1}{2\pi \mathrm{j}} \int_{\mathrm{L}} \exp(tz - t^3/3)\,\mathrm{d}t = \frac{1}{\pi} \int_0^{\infty} \cos(sz + s^3/3)\mathrm{d}s$$

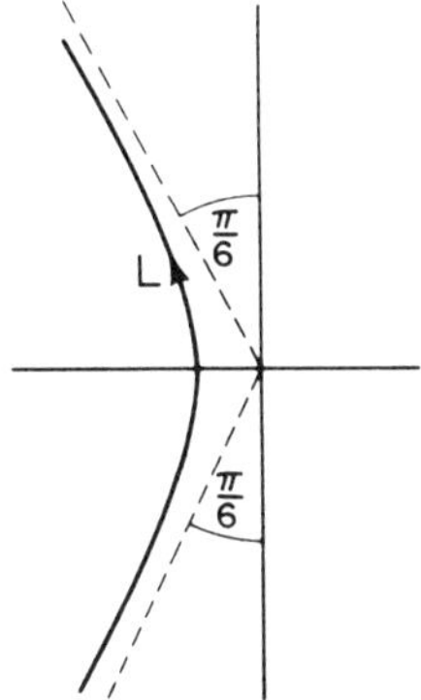

Airy's Integral

where L is a contour as shown in the diagram. It is the integral solution of the differential equation

$$\frac{\mathrm{d}^2 y}{\mathrm{d}z^2} - zy = 0$$

Airy's Points
The optimum points at which a bar must be suspended horizontally to make bending a minimum. The distance apart of the points is

$$\frac{l}{\sqrt{(n^2 - 1)}}$$

where l is the length of the bar and n the number of supports.

Airy's Spirals

The spirals of light which can be produced when convergent polarized light is passed through two plates cut from right-handed and left-handed quartz (enantiomorphic forms of the crystal) and observed with crossed Nicol prisms (i.e. prisms to separate the ordinary and extraordinary rays). The spirals arise from the rotation of the plane of polarization of the light and the direction of spiralling depends on which quartz plate is positioned first.

Aitken's Formula

If a series of numbers $u_1, u_2, \ldots u_n, u_{n+1}, u_{n+2}, \ldots$, is expected to converge slowly to an unknown limit u, then u can be estimated by assuming

$$u_n = u = e; \quad u_{n+1} = u - ke; \quad u_{n+2} = u - k^2 e \quad (k < 1)$$

whence, by eliminating k and e:

$$u = \frac{u_{n+2}u_n - u_{n+1}^2}{u_{n+2} - 2u_{n+1} + u_n} = u_{n+2} - \frac{(u_{n+2} - u_{n+1})^2}{u_{n+2} - 2u_{n+1} + u_n}$$

Alfvén Waves

In 1942 Alfvén predicted the possible existence of magnetohydrodynamic waves – in the simplest case transverse oscillations of magnetic field lines carrying with them a surrounding non-viscous perfectly conducting fluid. The velocity of propagation usually approximates (in SI) to

$$v^2 = \frac{B^2}{\mu\rho}$$

where B is the magnetic induction, ρ the density of the fluid and μ the magnetic permeability.

Amagat *See* **Appendix**

Amagat–Leduc Rule

According to E. H. Amagat and A. Leduc the volume occupied by a mixture of gases is equal to the sum of the volumes that the constituent gases would individually occupy at the temperature and pressure of the mixture. The Amagat–Leduc rule and **Dalton's Law of Partial Pressures** are identical for the perfect gas.

Amonton's Law
In any gas whose volume and mass are kept constant, the same rise in temperature produces the same increase of pressure.

Ampere; Ampere, Thermal; Ampere-turn *See* **Appendix**

Ampère's Law
The magnetic induction B produced at a point P in free space by a current flowing in a conductor is given using SI units by

$$B = \int \frac{\mu_0 I \, \mathrm{d}s \sin\theta}{4\pi r^2}$$

where μ_0 is the permeability of free space, I is the current flowing in an element of the circuit ds, r is the length of the line joining ds and the point P, and θ is the angle between ds and r. The law is also called **Laplace's Law** or the **Biot–Savart Relation.**

Alternatively, Ampère's Law is sometimes stated as

$$\oint B \cos\theta \, \mathrm{d}\ell = \mu_0 I$$

where the integral is over a closed path enclosing a current I which sets up magnetic induction B. θ is here the angle between B and the incremental path length dℓ.

Anderson Localization
For understanding the theory of electrons in certain non-crystalline media, P. W. Anderson has proposed the use of a potential-energy function which is non-periodic by the addition of a random potential energy to the periodic function used for crystalline solids. If the ratio of this random potential energy to the energy bandwidth associated with the original periodic function is higher than approximately 5 (the ratio depends slightly on the coordination number associated with the material), the wave-functions for the electrons decay exponentially with distance, thus giving rise to electron localization.

Andrade's Creep Law
Andrade showed that when a load is applied at the beginning of a creep test, the instantaneous elastic elongation is followed by a transient state in which strain varies as $t^{1/3}$ and, finally, a steady state is reached in which there is a constant rate of creep under constant effective stress.

Ångström *See* **Appendix**

Ångström's Formula
For the scattering effect of dust in the atmosphere

$$S = A\lambda^{-B}$$

where λ is the wavelength, B depends on the particle size and A is a constant.

Angus–Smith Process
When iron is heated to about 370°C and then immersed in a solution of coal-tar in oil and paranaphthalene, an anti-corrosive layer is formed. *See* **Bower–Barff Process**.

Antoine Equation
An expression for the vapour pressure p of a condensed solid or liquid as a function of absolute temperature T:

$$\ln p = A - \frac{B}{C+T}$$

where A, B and C are constants obtained experimentally.

Antonoff's Rule
The interfacial surface tension γ_{AB} between two saturated liquid layers A and B, in equilibrium, is equal to the difference between the surface tensions against vapour or air of the two mutually saturated solutions, i.e.

$$\gamma_{AB} = \gamma_A - \gamma_B$$

Apollonius' Circle
The locus of the vertex, A, of a triangle of given base BC such that the sides AB and AC are in a given ratio λ:1 is a circle with, as diameter, the line joining the points which divide the base BC internally and externally in the ratio λ:1.

Apollonius' Conic Sections
The curves obtained by cutting through a cone at particular angles. The ellipse, parabola and hyperbola were named by Apollonius.

Apollonius' Theorem
In any triangle, the sum of the squares on two sides is equal to twice the square on half the base together with twice the square on the median drawn to the base.

Appleton Layer *See* **Heaviside Layer**

Arbusov Reaction (or **Rearrangement**)
In the Arbusov rearrangement, an alkyl phosphite is heated with a small amount of the corresponding alkyl halide. Initially a phosphonium salt is formed which decomposes to give a dialkyl alkyl phosphonate. The formation of the phosphonium salt is so fast that, if an equivalent amount of a different alkyl halide is used, up to 95 per cent yield of a product having the new alkyl group attached to the phosphorus can be obtained.

Archimedean Polyhedra
These are semi-regular polyhedra, also called Archimedean solids, for which every face is a regular polygon, although the faces are not all of the same kind. The faces must be arranged in the same order around each vertex. There are 13 Archimedean solids although two of these 13 can exist in enantiomorphic forms: i.e. in right-handed and left-handed forms. They can all be described in a sphere. Additionally there are 13 Archimedean duals in which all the faces are congruent and their polyhedral angles regular but not identical. Such solids are referred to as vertically-regular and have an inscribed sphere. Each Archimedean dual can be obtained from a corresponding Archimedean solid by replacing every vertex of the latter by the tangent plane to the sphere in which it can be described. *See also* **Platonic Polyhedra**.

Archimedes' Axiom
If a and b are any two positive rationals, an integer n exists, such that $nb > a$. Alternatively, it is called **Eudoxus' Theorem**.

Archimedes' Principle
When a body is immersed in a fluid, it experiences a buoyant force which manifests itself as an apparent loss of weight equal to the weight of fluid displaced.

Archimedes' Spiral

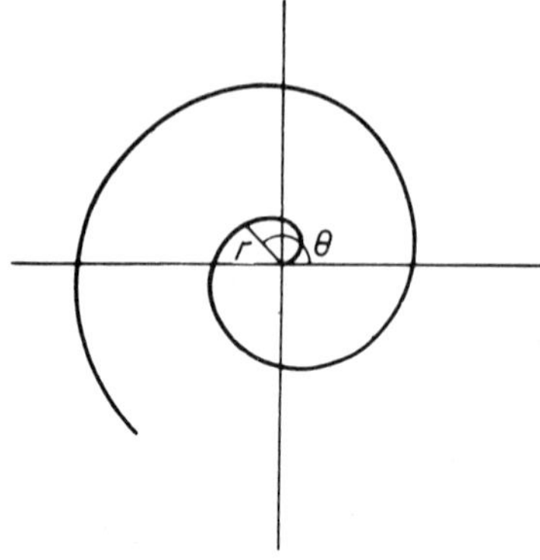

Archimedes' Spiral

A point moving uniformly along a line which rotates uniformly about a fixed point describes a spiral of Archimedes. The equation is: $r = a\theta$.

Argand Diagram

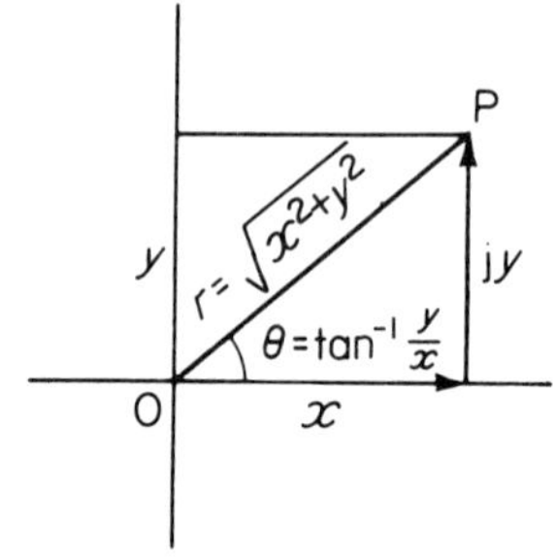

Argand Diagram

The representation of a complex quantity

$$z = x + jy$$

where x is the real and y the imaginary part, by a point P whose Cartesian coordinates are x, y. If z is expressed in De Moivre's form

$$z = re^{j\theta} = r(\cos\theta + j\sin\theta)$$

$r = \sqrt{(x^2 + y^2)}$, the length of the line from the origin O to P, and $\theta = \tan^{-1}(y/x)$ where θ is the angle between OP and the x axis.

Armstrong–Baeyer Benzene Formula *See* **Kekulé Benzene Formula**

Arndt–Eistert Synthesis
An aliphatic or aromatic acid may be converted to the next higher homologue by means of diazomethane. The acid chloride is treated with two moles of diazomethane and the diazoketone is warmed with water in the presence of a silver catalyst.

Arrhenius' Equation (for Reaction Velocity)
The most satisfactory method of expressing the effect of temperature on reaction velocity is due to S. Arrhenius and may be expressed by the equation

$$k = ve^{-E/RT}$$

where k is the specific reaction rate, E is the activation energy of the reaction, and v is the frequency factor.

Arrhenius' Theory
When an acid, base or salt is dissolved in a polar solvent a certain proportion of the molecules become spontaneously ionized.

Aston Dark Space
In a gas discharge the dark space in the immediate vicinity of the cathode, in which the emitted electrons have a velocity insufficient to excite the gas. *See also* **Faraday Dark Space**.

Aston Rule
The atomic weights of isotopes are approximately integers, and deviations of the atomic weights of the elements from integers are due to the presence of several isotopes with differing weights.

Atkinson Cycle
A working cycle for an internal combustion engine where the expansion ratio exceeds the compression ratio. It is more efficient theoretically than the **Otto Cycle** but is difficult to achieve in practice.

Auger Effect
When a K electron is ejected from an atom, an outer electron can fall into the K shell and give up its energy in two ways:

I. By radiation which gives rise to x-ray emission lines.

II. By transferring its energy to one of the outer electrons which is ejected from the atom. This is known as the Auger effect.

Avogadro's Law
Equal volumes of all gases, under the same conditions of temperature and pressure, contain equal numbers of molecules.

Avogadro's Number *See* **Appendix**

Azbel–Kaner Resonance
A type of cyclotron resonance in high-purity metals at liquid helium temperatures, first observed by Azbel and Kaner in 1957. A magnetic field is applied parallel to the surface of the metal in the same direction as an oscillating electric field. Current can only flow in a region next to the surface having a depth called the 'anomalous skin depth'. The radius of the electron orbits under the above conditions is much larger than this skin depth, so electrons can only be accelerated once per cycle and only if their orbits enter the skin depth region. The frequency of the electric field must be the cyclotron frequency or an integral multiple of it.

Babinet's Principle
The diffraction patterns obtained from a complementary pair of diffraction screens – that is two screens in which the opaque parts of one are replaced by transparent parts in the other – are the same.

Babo's Law
The lowering of the vapour pressure of a solvent by the addition of a non-volatile solute is proportional to the concentration of solute.

Back–Goudsmit Effect *See* **Paschen–Back Effect**

Badger Rule
An empirical relationship between the force constants and the vibrational frequencies of electron orbits in diatomic molecules.

Baeyer Strain Theory
The angle subtended by the corners and centre of a regular tetrahedron (109° 28′) lies between the values for the angle of a regular pentagon (108°) and a regular hexagon (120°). According to Baeyer therefore, *cyclo*paraffins in which the carbon atoms are co-planar are only possible when the minimum strain is set up between the carbon bonds, and for this reason the 5- and 6-membered rings should be most stable, as is found in practice. *See* **Sachse–Mohr Theory**.

Baeyer–Villiger Rearrangement
The aromatic cycloalkyl and aliphatic ketones with the carbonyl group attached to at least one secondary carbon atom can be oxidized using peracid to produce the ester. Cyclic ketones on the other hand give rise to the lactones.

$$\text{C}_6\text{H}_5\text{COCH}_3 \xrightarrow{\text{peracid}} \text{C}_6\text{H}_5\text{OCOCH}_3$$

$$\text{cyclopentanone} \xrightarrow{\text{peracid}} \text{δ-valerolactone}$$

Baily's Beads
Just before a solar eclipse becomes total, the advancing dark limb of the moon appears to break up into a series of bright points.

Baker–Nathan Effect (Hyperconjugation)
According to these workers the methyl group in propylene and certain other compounds is more electron-releasing than would be normally expected by the inductive effect. The phenomenon is only exhibited when the methyl group is attached to an unsaturated carbon atom.

Balmer *See* **Appendix**

Balmer Series
A series of lines observed in the atomic spectrum of hydrogen. The wavenumber of the lines of a spectral series may be represented by the equation

$$\bar{\nu} = R\left(\frac{1}{n_1{}^2} - \frac{1}{n_2{}^2}\right)$$

where R is the **Rydberg Constant**. Various values of n_1 and n_2 give a number of spectral series:

Series	n_1	n_2
Lyman	1	2, 3, 4, . . .
Balmer	2	3, 4, 5, . . .
Paschen	3	4, 5, 6, . . .
Brackett	4	5, 6, 7, . . .
Pfund	5	6, 7, 8, . . .

(*See also* **Ritz Combination Principle** and **Bohr's Theory**.)

Balz–Schiemann Reaction
Fluorobenzenes may be prepared by this reaction. When borofluoric acid is added to a diazonium salt the insoluble diazonium borofluoride is precipitated, collected by filtration, dried and heated gently when it is decomposed into fluorobenzene, nitrogen and boron trifluoride.

Bamberger's Formula
A structural formula for naphthalene in which the valencies of the benzene rings point towards the centres. The formula does not involve double bonds as in the Kekulé formula and bears no relation to the real structure of naphthalene which is a resonance hybrid.

Bardeen–Cooper–Schrieffer (BCS) Theory of Superconductivity
A quantum theory of superconductivity in which interaction between pairs of electrons gives, for the electron system, a ground state which is separated from the excited states by an energy gap. The presence of this energy gap accounts for the normal/superconducting phase transition and the critical magnetic field which leads to the destruction of superconductivity. The theory accounts for the **London Penetration Depth** and the **Pippard Coherence Length**.

Barkhausen Effect
When a ferromagnetic material is magnetized in an increasing magnetic field, the magnetization does not increase smoothly with field. This is because of the irregular fluctuations in the motion of the domain walls which separate regions of differing directions of magnetization in the material.

Barkhausen–Kurtz Oscillations
Oscillations occurring in an electronic valve, wherein electrons of high velocity are expelled from the filament, meet the repelling effect of the grid–anode field, stop, return through the grid, and so on, eventually landing on the grid wires. The current accompanying this motion may be used to produce oscillations of frequencies between 300 and 1500 MHz.

Barnett Effect
The magnetization acquired by an initially non-magnetized and stationary body when the body is rotated. This rotation will produce

the same intensity of magnetization in the body as a uniform axial field of magnetic intensity $H = \gamma\Omega$ where Ω is the angular velocity of rotation and γ the gyromagnetic ratio (the ratio of the angular momentum to the magnetic moment) of the elementary carrier. The effect has been measured in ferromagnetic materials.

The converse effect, the rotation of a body of magnetic material when a magnetic field is applied parallel to its axis, is called the **Einstein–de Haas Effect** or the **Richardson Effect**.

Bartlett Force
An exchange force between nucleons in which spin (equivalent to charge plus position) is exchanged. (*See also* **Majorana Force**.)

Bart Reaction
When a diazonium salt is treated with sodium arsenite in the presence of copper, the diazonium group is replaced by an arsenic group.

Bateman Equations
Equations applicable to a chain of radioactive nuclides denoted by subscripts 1 to n. The nth nuclide decays into the $(n+1)$th nuclide with a disintegration constant λ_n, and at any time t there are N_n atoms of the nth nuclide. Then

$$\frac{\mathrm{d}N_1}{\mathrm{d}t} = -\lambda_1 N_1$$

and

$$\frac{\mathrm{d}N_n}{\mathrm{d}t} = \lambda_{n-1} N_{n-1} - \lambda_n N_n.$$

If at $t = 0$ only the parent substance is present, and $N_1 = N_1{}^0$, $N_2{}^0 = N_3{}^0 = \ldots = N_n{}^0 = 0$, then the number of atoms of the nth member of the chain at time t is

$$N_n(t) = \sum_1^n \frac{\lambda_1 \lambda_2 \ldots \lambda_{n-1} N_1{}^0 \mathrm{e}^{-\lambda_n t}}{(\lambda_1 - \lambda_n)(\lambda_2 - \lambda_n) \ldots (\lambda_{n-1} - \lambda_n)}.$$

Baud *See* **Appendix**

Baudisch Reaction
o-Nitrosophenol may be prepared from benzene, by the action of nitrosyl radical in the presence of an oxidizing agent. The nitroso-

radical may be prepared by the reduction of nitrous acid or the oxidation of hydroxylamine in the presence of a copper salt which both stabilizes the free radical and prevents the formation of *p*-nitrosophenol.

Baumé Scale

A hydrometer scale in which 0° represents the specific gravity of water at 12·5°C and 10° the specific gravity of a 10 per cent solution of sodium chloride at the same temperature. Also known as the **Lunge Scale**.

Bauschinger Effect

If, at a temperature at which work-hardening occurs, a metal is deformed plastically under tension and then unloaded, the mechanical properties of the metal for further tensile loading are found to be different from those for compressive loading. The magnitude of the effect usually depends on the size of the tensile stress applied before the compressive loading.

Bayes' Theorem

If B is an arbitrary event whose probability $P(B)$ is not zero, and if A_1, $A_2, \ldots A_n$ are a series of mutually exclusive and exhaustive events, then the probability $P(A_j|B)$ of event A_j happening, once B has happened, is given by

$$P(A_j|B) = \frac{P(A_j)P(B|A_j)}{\sum_{i=1}^{n} P(B|A_i)P(A_i)}.$$

Thus the conditional probability of A_j happening, given B, is proportional to the unconditional probability of A_j multiplied by the conditional probability of B, given A_j.

Beattie–Bridgeman Equation of State

This is an equation with five adjustable constants which can be made to represent the behaviour of a real gas above the triple point:

$$p = \frac{RT(1-\varepsilon)}{V^2}.(V+B) - \frac{A}{V^2}$$

where p is the pressure of the gas, V is the volume, and

$$A = A_0\left(1-\frac{a}{V}\right); \qquad B = B_0\left(1-\frac{b}{V}\right); \qquad \varepsilon = \frac{c}{VT^3}$$

Becke Line

When two adjacent substances, for instance a crystal immersed in a liquid, are viewed under a microscope, then, if their refractive indices are close in magnitude, a narrow but bright line of light appears at the junction. The line is called the Becke line and arises from the combination of reflected and refracted light rays. If it is the crystal which has the higher refractive index, then as the objective of the microscope is raised, the Becke line appears to move into the crystal.

Beckmann Rearrangement

When treated with certain reagents, such as sulphuric acid, hydrochloric acid or phosphorus pentachloride, ketoximes undergo rearrangement to the acid amide. As this reaction is specific for anti-ketoximes, it may be used to determine the configuration.

Becquerel Effect

If two similar electrodes are immersed in an electrolyte, a current flows when one of the electrodes is illuminated.

Becquerel Rays

The radiation discovered by H. Becquerel (1896) to be emitted by uranium. Although uranium emits alpha, beta and gamma rays, they were beta rays which Becquerel detected using a photographic plate.

Beer's Law

If I is the intensity of the light transmitted through a solution at a given wavelength and I_0 is the intensity of the incident light, then, according to Beer

$$I = I_0\,10^{-\mu c d}$$

where c is the concentration of solution in gram-molecules per litre and d is the thickness of solution through which the light is transmitted. μ is known as the molar extinction coefficient of the solute for the particular wavelength. The Law is sometimes used in the form

$$I = I_0 \exp(-acd)$$

where *a* is the molar absorption coefficient.

Beilby Layer
A microcrystalline or amorphous layer formed on the surface of metals by polishing.

Beilstein's Test
This test detects the presence of halogens in many organic compounds. The organic halogen compound is heated in contact with cupric oxide (usually in the form of heated copper gauze) in a Bunsen flame. The volatile copper halide which is formed colours the flame green or blue. The test is not specific, as certain compounds (derivatives of pyridine, quinoline and purines for example) also give a positive result due to the formation of copper cyanide.

Bel *See* **Appendix**

Bergmann Method (of Polypeptide Synthesis)
During the preparation of polypeptides of known structure, it is necessary to control the position in which the condensation of a second amino acid can occur and, in general, to prevent ring closure. Many methods have been advanced to block the amino group of an amino acid during the preparation of a polypeptide, but that due to Bergmann is of most utility. The blocking agent in this case is benzyl oxycarbonylchloride, $C_6H_5CH_2O.CO.Cl$. The synthesis may be exemplified by

$$C_6H_5CH_2O.CO.Cl + H.N.H.CH_2.COOH \xrightarrow{\text{glycine}} \begin{matrix} CH_2.COOH \\ | \\ NH.OC.O.CH_2C_6H_5 \end{matrix}$$

$$\downarrow PCl_5$$

$$\begin{matrix} CH_2.CO.NH.CH_2.COOH \\ | \\ NH.OC.O.CH_2C_6H_5 \end{matrix} \xleftarrow{\text{glycine}} \begin{matrix} CH_2COCl \\ | \\ NH.OC.O.CH_2C_6H_5 \end{matrix}$$

$$\downarrow H_2 \text{ palladium black}$$

$$\underset{\text{glycyl glycine}}{\begin{matrix} CH_2.CO.NH.CH_2.COOH \\ | \\ NH_2 \end{matrix}} + CO_2 + CH_3C_6H_5$$

Bergmann Series
A series of spectral lines for alkalis lying in the infrared region. The series is sometimes called the fundamental series and the wave-numbers are given by

$$\bar{\nu} = R\left(\frac{1}{(3+d)^2} - \frac{1}{(m+f)^2}\right)$$

where R is the Rydberg constant; m takes the integral values 4, 5, ... The series of spectral lines for the alkalis is analogous to that for hydrogen (*see* **Balmer Series**) but the alkalis require the Rydberg corrections d and f in the equation.

Bernoulli–Euler Law
In a homogeneous bar, the curvature of the central fibre is proportional to the bending movement.

Bernoulli Polynomials
Bernoulli polynomials $B_r^{(n)}(x)$ of order n are given by the identity

$$\frac{t^n e^{xt}}{(e^t - 1)^n} = \sum_{r=0}^{\infty} \frac{t^r}{r!} B_r^{(n)}(x)$$

where $n = 0, 1, 2, \ldots$. In addition, the polynomials $B_r^{(n)}(0) = B_r^n$ are called Bernoulli numbers of order n, except for $n = 1$ when they are usually simply called **Bernoulli numbers**, B_r.

Bernoulli's Differential Equation
Bernoulli's differential equation is of the form

$$\frac{dy}{dx} + f(x)y = g(x)y^n$$

This equation may be made linear by the substitution

$$y = u^{1/(1-n)}$$

Bernoulli's Equation (of Liquid Motion)
Liquid escapes through an orifice of cross-sectional area a with a velocity v given by

$$v^2 = 2gh\frac{A^2}{A^2 - a^2}$$

where A is the surface area, h is the head of liquid and g is the acceleration due to gravity. For large values of A this equation reduces to **Torricelli's Law of Efflux**.

Bernoulli's Inequality
If $x > 0$, $y > 0$ and $n \neq 0$ or 1, then
(1) $x^n - 1 > n(x-1)$ if $x \neq 1$
(2) $nx^{n-1}(x-y) > x^n - y^n > ny^{n-1}(x-y)$ if $x \neq y$
unless $0 < n < 1$ when the inequality sign is reversed in both cases.

Bernoulli's Lemniscate
This is a special case of **Cassini's Ovals** (with $a = c$) and was so named by Bernoulli because its shape is like a pendant ribbon (Latin: *lemniscus*) or figure of eight. The polar equation for the curve (with pole at the centre) is of the form

$$r^2 = 2a^2 \cos 2\theta.$$

Alternatively, the Cartesian equation is

$$(x^2 + y^2)^2 = 2a^2(x^2 - y^2)$$

Bernoulli's Theorem (for Probability)
A sequence of independent trials, in which the probability of 'success' is the same for each, is termed a series of **Bernoulli trials.** If s is the number of 'successes' in a series of n independent trials and p is the probability of an individual 'success', then Bernoulli's theorem states that the probability P of the relative frequency of 'successes' ($= s/n$) differing from p by more than ε tends to zero as $n \to \infty$ for any $\varepsilon > 0$; i.e.

$$\lim_{n \to \infty} \mathrm{P}\left\{ \left| \frac{s}{n} - p \right| > \varepsilon \right\} = 0 \qquad (\varepsilon > 0)$$

Bernoulli's Theorem in a Field of Flow
At every point in a steadily flowing fluid the sum of the pressure head, the velocity head and the height is constant.

Berthelot Equation
The Berthelot equation is a modified gas equation for real gases of the form

$$pV = RT\left\{1 + \frac{9}{128} \cdot \frac{pT_c}{p_cT}\left(1 - 6\frac{T_c^2}{T^2}\right)\right\}$$

where p_c and T_c are the critical pressure and critical temperature of the gas. The equation holds best at low pressures and is applicable only over a limited range.

Berthelot–Thomsen Principle
In 1867, M. Berthelot postulated that, of all possible chemical reactions, the one accompanied by the greatest development of heat will occur. Changes of state must be excluded from this rule and there are other obvious deviations.

Bertrand's Test (for Convergence)

If $\quad \mathrm{f}(x) = \mathrm{O}(x^{\lambda-1}) \quad$ or if $\quad \mathrm{f}(x) = \mathrm{O}(x^{-1}[\ln x]^{\lambda-1})$

then $\quad \displaystyle\int_a^\infty \mathrm{f}(x)\,\mathrm{d}x$ converges when $\lambda < 0$.

Bessel–Clifford Differential Equation

$$u\frac{\mathrm{d}^2y}{\mathrm{d}u^2} + (n+1)\frac{\mathrm{d}y}{\mathrm{d}u} + y = 0$$

reduces to Bessel's equation (*see* **Bessel Functions**) with the substitution

$$y = zu^{-n/2} \quad \text{and} \quad u = x^2/4$$

The equation has a solution of the form

$$y = u^{-n/2}\,\mathrm{J}_n(2\sqrt{u}) + u^{-n/2}\,\mathrm{N}_n(2\sqrt{u}).$$

Bessel Functions
Bessel's equation

$$\frac{\mathrm{d}^2z}{\mathrm{d}x^2} + \frac{1}{x}\frac{\mathrm{d}z}{\mathrm{d}x} + \left(1 - \frac{n^2}{x^2}\right)z = 0$$

has as its general solution

$$z = A\mathrm{J}_n(x) + B\mathrm{N}_n(x)$$

where $\mathrm{J}_n(x)$ and $\mathrm{N}_n(x)$ are Bessel functions of order n''. $\mathrm{J}_n(x)$ are Bessel functions of the first kind:

$$J_n(x) = \frac{x^n}{2^n\Gamma(n+1)}\left\{1 - \frac{x^2}{2(n+1)} + \frac{x^4}{2^4 2!(n+1)(n+2)} - \frac{x^6}{2^6 3!(n+1)(n+2)(n+3)} + \ldots\right\}$$

$N_n(x)$ are Bessel functions of the second kind:

$$N_n(x) = \frac{2}{\pi} J_n(x)\left\{\ln\frac{x}{2} + \gamma\right\} - \frac{1}{\pi}\left(\frac{x}{2}\right)^{-n} \sum_{k=0}^{k=n-1} \frac{(n-k-1)!}{k!}\left(\frac{x}{2}\right)^{2k}$$

$$-\frac{1}{\pi}\left(\frac{x}{2}\right)^n \sum_{k=0}^{k=\infty} \frac{(-1)^k}{(n+k)!k!}\left(1 + \tfrac{1}{2} + \tfrac{1}{3} + \ldots\right.$$

$$\left.\ldots + \frac{1}{k} + 1 + \tfrac{1}{2} + \tfrac{1}{3} + \ldots \frac{1}{n+k}\right)\left(\frac{x}{2}\right)^{2k}$$

γ is **Euler's Constant**. $N_n(x)$ are sometimes referred to as **Neumann's** *or* **Weber's Bessel Functions of the second kind**.

Bessel Inequality

$$|A| + |B| \geqslant |A + B|$$

where A and B are two vectors.

Bessel's Integral Equation

A **Bessel Function** of nth order can be represented by the integral equation

$$J_n(x) = \frac{1}{\pi}\int_0^{\pi} \cos(nt - x\sin t)\,dt \qquad (n = 0, 1, 2, \ldots)$$

Bessemer Process

A process used in the manufacture of steel whereby the impurities present in pig iron are directly oxidized by a blast of air. The process depends upon the oxidation of carbon to carbon monoxide and silicon to silica which reacts with the metal to form a silicate slag. Phosphorus impurities are more difficult to deal with and in the original Bessemer process where the lining of the converter was ganister, a siliceous material, a phosphorus-bearing iron could not be used. However, if the converter is lined with calcium oxide and magnesium oxide bound together with tar and lime added at the end of the blow, the phosphorus can be almost completely removed. This modification is the basis of the **Gilchrist–Thomas Process**.

Betti's Reciprocal Theorem
If an elastic body is subjected to two systems of body and surface forces, then the work that would be done by the first system of forces in acting through the displacements due to the second system of forces is equal to the work that would be done by the second system of forces acting through the displacements due to the first system of forces.

Betts' Process
An electrolytic process for refining lead. The electrolyte is a solution of lead silicofluoride and hydrofluosilicic acid. The lead is deposited on the cathode and the impurities remain on the anode.

Bhabba Scattering
Positron scattering by an electron.

Bienaymé–Chebyshev Inequality
If $x_1, x_2, \ldots x_n$ are uncorrelated random variables which have a mean value $\bar{x}\left(=\frac{1}{n}\sum_{i=1}^{n} x_i\right)$ and which are from a distribution having a mean value m and standard deviation σ, then for any $\lambda > 0$,

$$\mathrm{P}\left\{(\bar{x}-m) \geqslant \frac{\lambda\sigma}{\sqrt{n}}\right\} \leqslant \frac{1}{\lambda^2}$$

where P represents probability.

Bingham Fluid
A substance which remains rigid under a shear stress until the magnitude of the stress exceeds the yield stress, whereupon the substance flows like a **Newtonian Liquid**.

Biot *See* **Appendix**

Biot–Fourier Equation
An equation for the conduction of heat through a solid:

$$\frac{\partial Q}{\partial t} = \frac{\lambda}{C\rho}\nabla^2 T$$

where $\partial Q/\partial t$ is rate of flow of heat, λ the thermal conductivity and C the specific heat of the solid of density ρ. (*See also* **Fourier's Law**.)

Biot Number *See* **Nusselt Number**

Biot–Savart Relation *See* **Ampère's Law**

Biot's Law

The rotation of the plane of polarization of light propagated through an optically active medium is proportional to the length of the path, to the concentration (if the medium is a solution) and approximately inversely proportional to the square of the wavelength of the light.

Birch Reduction

A method of producing non-benzoid cyclic ketones by reduction of a phenol with sodium in liquid ammonia in the presence of ethyl alcohol.

Birkeland–Eyde Method

The fixation of nitrogen from the air depends upon an endothermic reaction between oxygen and nitrogen and is therefore favoured by high temperatures. As the reaction is reversible, rapid cooling of the product, nitric oxide, is necessary in order to prevent dissociation. The method of Birkeland and Eyde, although of little practical importance today, was the first method used commercially and depended upon passing the air through an electric arc.

Bischler–Napieralski Reaction

β-phenylamides can be dehydrated with ring closure to form 3, 4-dihydroisoquinoline derivatives by means of condensing agents such as phosphorus pentoxide or zinc chloride.

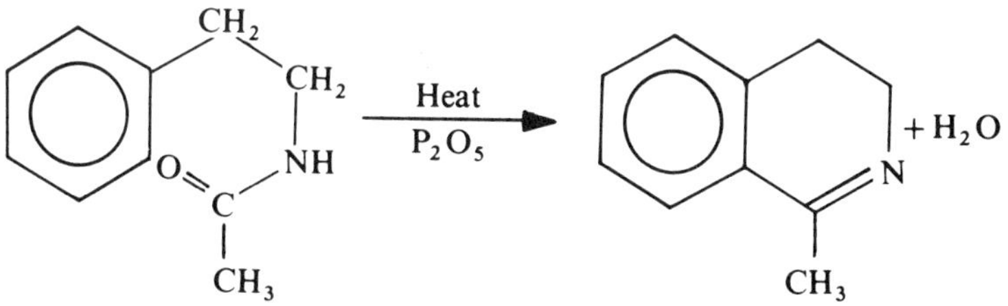

Bitter Figures
A method of viewing domains in a ferromagnetic material is to polish the material electrolytically to obtain a strain-free surface and then to spread a colloidal suspension of magnetite particles on the surface. The particles are drawn to regions of local maxima in the magnetization and a powder pattern or Bitter figure is obtained.

Bjerrum Theory of Molecular Spectra
In 1912 Bjerrum pointed out that the width of the infrared absorption bands of gases was of the order of magnitude to be expected on the assumption that they were due to molecular vibration in combination with molecular rotation. This doublet structure had already been observed in 1910 by A. Trowbridge and R. Wood but its significance had not been recognized.

Bjerrum's Treatment of Ion Association
In the calculation of the electric density in the vicinity of an ion in an electrolyte, Bjerrum suggested that certain difficulties associated with the integration of **Poisson's Equation** could be avoided by using the concept of the association of ions to form ion-pairs.

Blagden's Law
The fact that a dissolved substance depresses the freezing point of water has been known for many years. In 1771 R. Watson observed that the time a solution took to freeze was proportional to the concentration. That the depression of freezing point is proportional to the concentration of the dissolved substance was observed by C. Blagden in 1788 and F. Rudorf in 1861.

Blanc Rule
A generalization that states: whereas succinic and glutaric acids yield cyclic anhydrides on pyrolysis, adipic and pimelic acids yield cyclic ketones. It has been used to determine whether oxygenated or unsaturated rings of compounds of unknown composition are five- or six-membered. Thus, when a five-membered ring is oxidized it gives glutaric acid which pyrolyses to an anhydride, whereas a six-membered ring gives adipic acid and hence a ketone. The rule should be used with caution as certain cases are known where, for example, substituted adipic acid gives rise to a seven-membered cyclic anhydride.

Bloch's Functions
The solutions of the Schrödinger equation with a periodic potential are of the form

$$\psi = u_k(r)\,\mathrm{e}^{\mathrm{j}k.r}$$

where u is a function depending in general on k, which is periodic in x, y, z with the periodicity of the potential; that is, with the period of the lattice. This result was known earlier as **Floquet's Theorem**.

Bloch's Law
All flashes of light shorter than some critical duration cause equal effects on the eye, however many quanta are contained within a flash. The critical duration is approximately thirty milliseconds.

Bloch's $T^{3/2}$ Law for Magnetization
At very low temperatures ($T/T_C \ll 1$, where T_C is the **Curie Temperature**), the spontaneous magnetization M_S may be expressed as a function of temperature by

$$M_{S,T} = M_{S,0}(1 - CT^{3/2})$$

where C is a constant equal to

$$\frac{0{\cdot}0587}{aS}\left(\frac{k}{2SJ}\right)^{3/2}$$

Here, $a = 1, 2$ and 4 respectively for simple cubic, body-centred cubic and face-centred cubic structures, whose ions have spin S. J is an exchange integral relating the energy of interaction with the magnitude and direction of the spins of neighbouring ions.

This law is in quite good agreement with observation in the very low temperature region but at somewhat higher temperatures a T^2 term replaces the $T^{3/2}$ term.

Bloch's Theorem of Superconductivity
Bloch proved that the lowest state of a quantum-mechanical system in the absence of a magnetic field can carry no current.

Bloch Wall
A ferromagnetic material in zero magnetic field will be in equilibrium if it has minimum potential energy. This cannot be so if the crystal has a single domain of one direction of magnetization. So the crystal

consists of domains which are magnetized in different directions Bloch walls are the transition layers between domains which occur in bulk samples. The magnetic spin orientation of the atoms gradually changes within the layer, with the magnetic spin vector rotating about an axis perpendicular to the boundary. This arrangement is energetically more favourable than one in which the axis of rotation is in the plane of the boundary. For the latter case, free poles would be produced over the whole boundary wall and this would correspond to a larger potential energy. However, for thin film samples the second case is energetically more favourable and the wall then formed is called a **Néel Wall**.

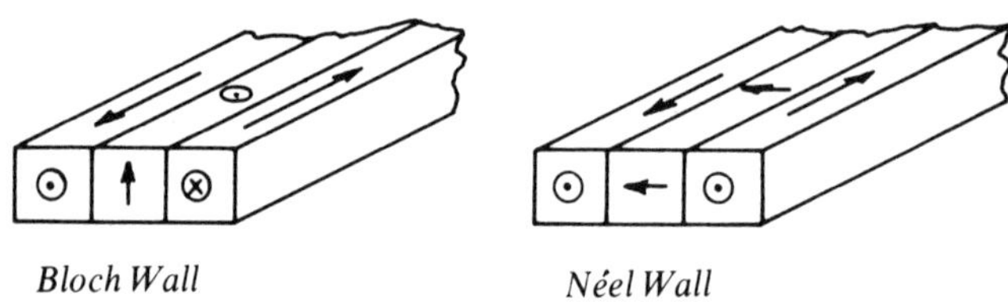

Bloch Wall *Néel Wall*

Blondel *See* **Appendix**

Blondel–Rey Law
A light source, flashing at low frequency (less than 5 hertz) and having brilliance B_0 during the flash, has an apparent brilliance B given by

$$B = \frac{B_0 t}{\alpha + t}$$

where t is the time of duration of the flash and α is a constant (equal to approximately 0·2 s when B is near the threshold for white light). *See also* **Talbot's Law**.

Bode's Law
Bode stated that, if the distance of Mercury from the Sun is taken as 4, the corresponding distances of the other planets would be given, in the same units, as: Venus 4 + 3, the Earth 4 + 6, Mars 4 + 12, Asteroids 4 + 24, Jupiter 4 + 48, Saturn 4 + 96 and Uranus 4 + 192. Neptune however, which should lie according to Bode's law at a distance of 388 from the Sun, was found to be in fact 300 units away. Sometimes called the **Titius–Bode law**.

Bodroux–Chichibabin Reaction
A method of synthesizing aldehydes in which reaction occurs between ethyl orthoformate and a Grignard reagent (*see* **Grignard Reaction**) containing the alkyl group required in the aldehyde. The acetal formed is hydrolysed in acid solution to give the aldehyde.

Boettger's Test
A test for the presence of saccharoses based on the reduction of bismuth nitrate to metallic bismuth in alkaline solution.

Bohr–Breit–Wigner Theory
A nuclear reaction occurs in two stages. In the first, the bombarding particle is absorbed by the target nucleus to form a high-energy unstable compound nucleus. In the second, this nucleus emits radiation to leave a residual lower-energy nucleus known as a recoil nucleus.

Bohr Magneton; Bohr Radius *See* **Appendix**

Bohr's Correspondence Principle
The behaviour of the electrons in an atomic system must approach more and more that predicted by classical physics the higher the quantum number of the orbit.

Bohr's Theory
Following Rutherford's postulate that atoms consist of a positively charged nucleus surrounded at a relatively large distance by a number of negatively charged electrons, N. Bohr developed in 1913 a theory which gave quantitative agreement with certain spectroscopic data. Bohr suggested that there were certain circular orbits (later modified to elliptical orbits by Sommerfeld) and that the energies of these were determined by the quantum condition that the angular momentum of the electrons is an integral multiple of $h/2\pi$. Electromagnetic radiation is emitted if an electron initially moving in an orbit of total energy E_i discontinuously changes its motion to move into an orbit of energy E_f, in which case the frequency of emission is equal to $(E_i - E_f)/h$ where h is **Planck's Constant**. Hence for hydrogen or positively charged helium, Bohr showed the emitted radiation to be

restricted to wavenumbers specified by

$$\tilde{\nu} = R\left(\frac{1}{T_2^{\,2}} - \frac{1}{T_1^{\,2}}\right)$$

where T_1 and T_2 are positive integers and R is the **Rydberg Constant**, such that

$$R = \frac{2\pi^2 e^4 Z^2 \mu}{h^3}$$

where Z is the nuclear charge and μ the reduced mass.

Bohr regarded absorption of light as the reverse of emission. Also resonance radiation is produced by the return of an electron to the same ground state from which it had been excited.

Bohr–Van Leeuwen Theorem
These workers have shown that, in any dynamical system to which classical statistical mechanics can be applied, the paramagnetic and diamagnetic susceptibilities cancel, so that, classically, for thermal equilibrium, one always has zero susceptibility.

Bohr–Wheeler Theory of Fission
A theory of fission based on a liquid-drop model of the nucleus in which there are coulombic and surface tension forces. (*See also* **Weizsäcker's Formula**.)

Boltzmann Distribution Law *See* **Maxwell–Boltzmann Distribution Law**

Boltzmann Equation
The entropy of a system of particles is proportional to the logarithm of the probability of its macroscopic state. The constant of proportionality is known as **Boltzmann's Constant**, (*See* **Appendix**.)

Boltzmann H Theorem
The **Boltzmann H function** is defined as

$$\mathrm{H} = \iint \mathrm{f} \ln \mathrm{f} \, d\mathbf{v} \, d\mathbf{r}$$

where integration is over all position and momentum space. f is a distribution function and **v** and **r** are velocity and position vectors respectively. Then, the Boltzmann H theorem states that H is a decreasing function with time unless f is Maxwellian in which case H remains constant. This is analogous to the thermodynamic statement that the entropy of a system tends to a maximum, and, except for an additive constant, the H function equals the negative of entropy divided by Boltzmann's constant.

Boltzmann's Law of Radiation (Stefan–Boltzmann Law)
Stefan induced an experimental law for the intensity of the total radiation from a black-body at various temperatures. This law was subsequently derived theoretically by Boltzmann. The law states that the complete emission W of a black-body is proportional to the fourth power of the absolute temperature.

$$W = aT^4$$

where a is a constant which includes **Stefan's Constant**. (*See* **Appendix**.)

Boltzmann's Ratio
Where particles can have two energy levels not widely separated, the distribution of the particles over these levels is controlled by Boltzmann's ratio

$$\frac{n_2}{n_1} = \exp\left(\frac{-\Delta E}{k_B T}\right)$$

where ΔE is the energy difference between the levels and n_2 particles are in the upper level and n_1 in the lower level.

Boltzmann Statistics *See* **Maxwell–Boltzmann Statistics**

Boltzmann's Transport Equation
An equation expressing the variation $(f-f_0)$ of a distribution function for charged carriers from its equilibrium value f_0 in the presence of a thermal gradient and fields such as electric and magnetic fields. If $\mathbf{v}$ is the velocity of the carriers, $\mathbf{p}$ their momentum and $\mathrm{d}\mathbf{p}/\mathrm{d}t$ the force on them at time t, and $\tau(\mathbf{p})$ is a relaxation time for the collision processes, then, for elastic scattering of the carriers and small perturbations, it can be written in the form

$$\frac{\partial f}{\partial t} + \mathbf{v}\nabla f + \frac{\mathrm{d}\mathbf{p}}{\mathrm{d}t}\nabla_p f = \frac{f-f_0}{\tau(\mathbf{p})}$$

For steady-state processes $\partial f/\partial t = 0$.

Bolzano–Weierstrass Theorem
A bounded sequence of numbers always contains a convergent sequence. (A sequence is bounded if there is a number A such that, for all values of n, $|x_n| \leqslant A$.) For instance, consider the bounded but non-convergent sequence $\{x_n\}$ where $x_n = (-1)^n$. This has a sub-sequence $\{x_{2n}\}$ which converges to 1.

Bolza, Problem of
The general problem of finding the class of curves $y_i = y_i(x)$ such that

$$\int_{a(z)}^{b(z)} \mathrm{F}_i(x, y, y')\,\mathrm{d}x + \mathrm{G}_i(z)$$

are minima subject to constraints of the form

$$\Phi_j(x, y, y') = 0.$$

For fixed points, $\mathrm{G}_i(z) = 0$, and the problem becomes an isoperimetric one; for example, the problem of finding the path between two points of a particle of pre-assigned initial velocity under constant acceleration due to gravity.

Bonnet's Form
If $\mathrm{f}(x)$ and $\phi(x)$ are integrable between a and b and if $\phi(x)$ is a positive decreasing function of x, then a number ξ exists such that $a \leqslant \xi \leqslant b$ and

$$\int_a^b \mathrm{f}(x)\phi(x)\,\mathrm{d}x = \phi(a)\int_a^\xi \mathrm{f}(x)\,\mathrm{d}x.$$

Boolean Algebra
An algebra of sets used to enunciate logical propositions and their consequences. The algebra has two binary operations called addition and multiplication. The algebra may be used to represent binary logic and can be applied to computer logic, switching networks and so on. *See also* **De Morgan's Rules** *and* **Venn Diagram**.

Boord Olefin Synthesis

Well-defined olefins can be formed from aldehydes by converting them to α-alkyl β-bromoalkyl ethers and introducing the necessary double bond by removing $C_2H_5O\,Zn\,Br$ by reaction with zinc.

$$2RCH_2CHO \xrightarrow[C_2H_5OH]{HCl} 2RCH_2\underset{\displaystyle Cl}{\underset{|}{C}}HOC_2H_5 \xrightarrow{Br_2} 2R\underset{\displaystyle Br}{\underset{|}{C}}H-\underset{\displaystyle Br}{\underset{|}{C}}HOC_2H_5 \xrightarrow{R'MgX}$$

$$2R\underset{\displaystyle Br}{\underset{|}{C}}H-\underset{\displaystyle R'}{\underset{|}{C}}HOC_2H_5 \xrightarrow{Zn} 2RCH=CHR'+C_2H_5OC_2H_5+Zn\,Br_2$$

Bordoni Effect

When metals possessing a close-packed crystal structure are subjected to an oscillating stress, there is a peak in the internal friction at a particular frequency of applied stress. It is considered to be due to dislocations, which run parallel to close-packed rows, oscillating from one equilibrium position to the adjacent equilibrium position. For frequencies of 1 kHz to 1 MHz, the maximum appears at approximately one third the Debye temperature (*see* **Debye Equation**).

Born Approximation

A method of obtaining the asymptotic form for a scattered wave (e.g. in quantum mechanics) depending on performing a sequence of iterative integrations.

Born–Haber Cycle

An imaginary cycle for a crystalline halide used to calculate its lattice energy. The steps in the cycle are:

(1) Heat is removed from the crystal to reduce its temperature to absolute zero.
(2) The crystal is broken into free stationary ions.
(3) Electrons are transferred from the negative ions to positive ions to produce neutral atoms.
(4) The halogen atoms are allowed to recombine to form diatomic atoms and
(5) the gas is brought to room temperature and pressure.
(6) The metal vapour is also returned to room temperature and

(7) condensed to the metal.
(8) The metal and the halogen are allowed to combine chemically, thus completing the cycle.

If the energy associated with each step except (2) is known, the lattice energy can be calculated.

Born and von Kármán Theory (for Heat Capacity of a Solid)
A theory of the heat capacity of a crystalline solid, based on the lattice dynamics of the atoms. The model assumes a particular approximate frequency spectrum applicable to a linear chain, with a cut-off number based on the normal modes of vibration of the atoms in the chain. The spectrum reduces to that used in the Debye model if an elastic medium is assumed. (*See also* **Debye Equation** for heat capacity of a solid.)

Borrmann Effect (or **Campbell–Borrmann Effect**)

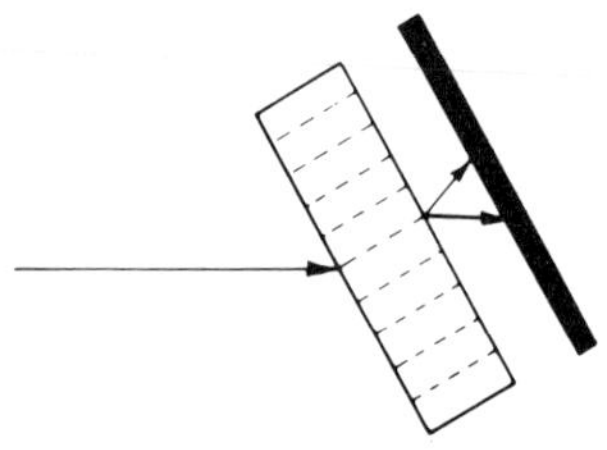

Borrmann Effect

The anomalous transmission of x-rays when a single crystal of high perfection is placed in a reflecting position in a monochromatic x-ray beam. The transmitted and diffracted beams emerge displaced and with anomalously high intensity. Absorption is reduced because standing waves are established in the crystal with nodes at the atomic sites and the displacement is because energy flow in the crystal is along the reflecting planes. The effect is the basis of an x-ray topography method of photographing dislocations in solids.

Bosanquet's Law
The resistance of an electric circuit is defined as the ratio of the

electromotive force to the current (**Ohm's Law**). In a magnetic circuit the ratio of the magnetomotive force F_m to the magnetic flux Φ is known as the reluctance R. Bosanquet's Law states that

$$\Phi = \frac{F_m}{R} \text{ in analogy to Ohm's Law.}$$

Bose–Einstein Condensation

As the temperature of an assembly of Bose–Einstein particles tends to zero, the particles are able to fall into the zero energy state. This is called Bose–Einstein condensation and is condensation into momentum space rather than condensation into coordinate space which occurs for liquefaction from the gas phase. The behaviour is in contrast to that of an assembly of Fermi–Dirac particles where only one particle per state including the zero energy state is allowed because of the **Pauli Exclusion Principle**, and where, at zero temperature, particles fill the states up to the Fermi energy level.

Bose–Einstein Statistics

The study of the probability of occupation of energy states by indistinguishable non-interacting independent particles in a quantized system. These particles are allowed to be arranged among the energy states without limit to the number of particles per state, and this fact differentiates Bose–Einstein statistics from **Fermi–Dirac Statistics**. The probability that an energy state E is occupied is given by the **Bose–Einstein distribution law**:

$$\mathrm{f}(E) = \{\exp[(E-\mu)/k_B T] - 1\}^{-1}$$

where μ is a constant. Particles which are subject to Bose–Einstein statistics are called Bose–Einstein particles or bosons.

Bouguer's Law *See* **Lambert's Law**

Bouvealt–Blanc Reduction

Alcohols may be prepared by refluxing an ester with sodium and an alcohol. The amides of the aryl fatty acids are reduced under the same conditions to the corresponding amine.

$$RCOOCH_3 + 4H \rightarrow RCH_2OH + CH_3OH$$

Bower–Barff Process
A method of reducing the corrosion of iron or steel wherein the meta is raised to a red heat and is then subjected to superheated steam.

Bow's Notation
A method of notation of forces acting at a point, wherein the spaces between forces are lettered or numbered, any force being defined by the letters or numbers on each side of it.

Boyle's Law (Marriotte's Law)
At a constant temperature the volume of a definite mass of gas is inversely proportional to the pressure, i.e.

$$pV = \text{constant.}$$

Real gases deviate appreciably in behaviour from this equation, as might be expected as the law may be deduced from kinetic theory, which considers the molecules of the gas as point masses having no attraction for one another.

Boyle's Temperature (Point)
In general, real gases do not obey **Boyle's Law**. The relation between the pressure and the volume of a real gas may be expressed by a power series

$$pV = A + Bp + Cp^2 \ldots$$

The coefficients A, B, C, etc., are called the virial coefficients. The first virial coefficient A is equal to RT where T is the absolute temperature and the effect of the third and higher virial coefficients is small except at high pressures. The deviation of the behaviour of the gas from Boyle's Law is therefore determined by the second virial coefficient B which is temperature-dependent. The temperature at which B becomes zero is known as the Boyle temperature, for, at this temperature, provided the higher virial coefficients are negligible, the gas will obey Boyle's Law up to comparatively high pressures.

Brackett Series *See* **Balmer Series**

Bragg–Gray Principle
The energy absorbed per unit mass $(\mathrm{d}E/\mathrm{d}m)_\mathrm{b}$ for a particular medium

b is related to the ionization which occurs in a small gas-filled cavity in that medium by

$$(\mathrm{d}E/\mathrm{d}m)_{\mathrm{b}} = P_{\mathrm{b}} W_{\mathrm{g}} J_{\mathrm{g}}$$

where P_{b} is the mass stopping power of the medium relative to the gas g, W_{g} is the average energy required to produce an ion-pair in a gas and J_{g} is the number of ion pairs produced per unit mass of gas in the cavity (determined experimentally). The principle has application to the measurement of ionizing radiation.

Bragg–Kleenan Rule
The atomic stopping power of a material to charged particles is proportional to the square root of the material's atomic weight. Atomic stopping power is additive and thus can be obtained for complex molecules.

Bragg–Pierce Law
The atomic absorption coefficient of an element for x-rays is given by

$$\mu_{\mathrm{a}} = CZ^4\lambda^3 \text{ (sometimes quoted } \mu_{\mathrm{a}} = C'Z^5\lambda^{7/2}\text{)}$$

where Z is the atomic number, λ is the wavelength of the x-rays and the constant C (or C') varies with λ.

Bragg's Law
Let the distance between lattice planes in a crystal be d, let the wavelength of a homogeneous beam of x-rays be λ and let the glancing angle be θ. Then for diffraction

$$n\lambda = 2d \sin\theta \qquad (n \text{ an integer})$$

This relation, known as the **Bragg equation**, enables the crystal parameters to be determined.

Bravais Lattice
A crystal is built up by the repetition by translation in three dimensions of a structural unit. The nature of the basic pattern is apparent if each of these structural units is replaced by a point. In this infinite three-dimensional array each point is in the same environment as any other point. Bravais showed that there can only be 14 of these space lattices. The unique lattices are shown overleaf.

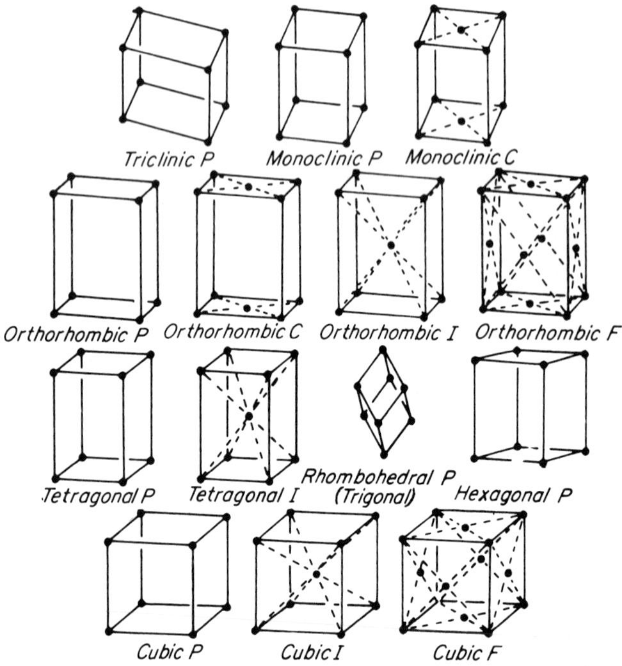

Bravais Lattices

Bravais' Law
Those crystal forms which tend to occur most frequently are those with faces parallel to planes of smallest reticular area. (Reticular area is inversely proportional to the interplanar spacing.) *See also the* **Donnay–Harker Principle.**

Bredt Rule
Reactions of the **Diels–Alder** type with cyclopentadiene give cyclanes having a methylene bridge across a cyclohexane ring. The corresponding alcohol does not hydrate, as double bonds cannot exist in the bridge.

Breit–Wigner Formula
If a nucleus is excited to a well-defined single excitation level by an incoming particle *n*, for instance a neutron, with the emission of a

particle p, then the reaction cross-section is given by

$$\frac{\pi g}{k^2} \frac{\Gamma_n \Gamma_p}{(E - E_r)^2 + (\Gamma/2)^2}$$

where
$$g = \frac{2J + 1}{(2I + 1)(2s + 1)}.$$

J is the spin of the compound nucleus, I the spin of the target nucleus and s the spin of the incoming neutron n. Γ_n and Γ_p are the neutron and particle energy widths and Γ is the total energy width of the compound nucleus (proportional to the disintegration constant). E_r is the energy at the peak of the resonance, E the total energy (of the target and incoming neutron) and k the wave-vector associated with the incident neutron.

Brewster *See* **Appendix**

Brewster's Fringes
Interference fringes observed when light from an extended source passes through two etalons (pairs of parallel semi-silvered glass plates with an air-gap between) which are inclined at a small angle. There is a path difference established between light waves reflected in the two etalons, which must be of equal thickness or whose ratio of thicknesses is an integer. Usually, transmission fringes are observed, but the reflection fringes in a Jamin interferometer are a particular case of Brewster's fringes.

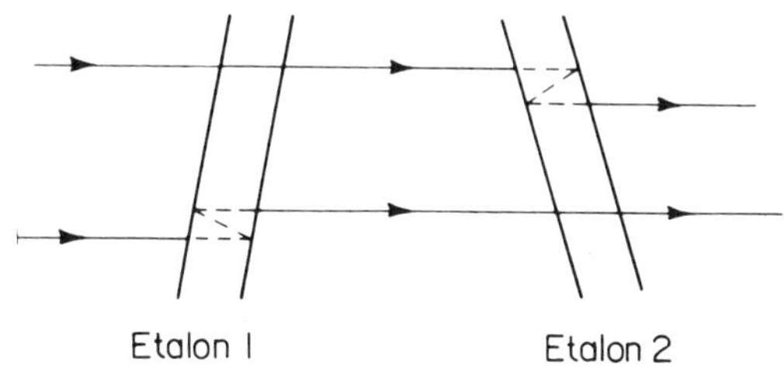

Brewster's Fringes

Brewster's Law
When a beam of light is reflected from a surface, polarization may occur. As the angle of incidence is increased from normal to grazing

the polarization increases, passes through a maximum and then decreases. The angle of maximum polarization is known as the angle of polarization or **Brewster angle** and varies with the nature of the substance. Brewster discovered that the refractive index was equal to the tangent of the angle of polarization. This law leads to the result that the sum of the angle of incidence and angle of refraction at maximum polarization is equal to a right angle.

Brianchon's Theorem
For any hexagon whose sides are tangent to a conic, the diagonals connecting opposite vertices intersect in a point (or are parallel).

Bridgman Effect
This effect is the absorption or liberation of heat when an electric current passes through an anisotropic crystal. It arises from non-uniformity in the current distribution and is additional to the **Peltier** and **Thomson Effects**.

Bridgman Relation
In a metal or semiconductor

$$PK = QT$$

where P is the Ettingshausen coefficient, K is the total thermal conductivity and Q is the Nernst–Ettingshausen coefficient. (*See* **Ettingshausen** and **Nernst Effects**.)

Bridgman–Stockbarger Method
A method of single-crystal growth in which a melt, contained in a capsule, is lowered through a temperature gradient about the freezing point. The method involves either a two-zone furnace, in which case the upper zone is at a temperature above, and the lower zone at a temperature below, the freezing point, or a single-zone furnace, when use is made of the drop in temperature down the axis of the furnace. By either tapering the capsule or constricting it at the lower end, a single crystal is selected. A horizontal modification of the method can also be used.

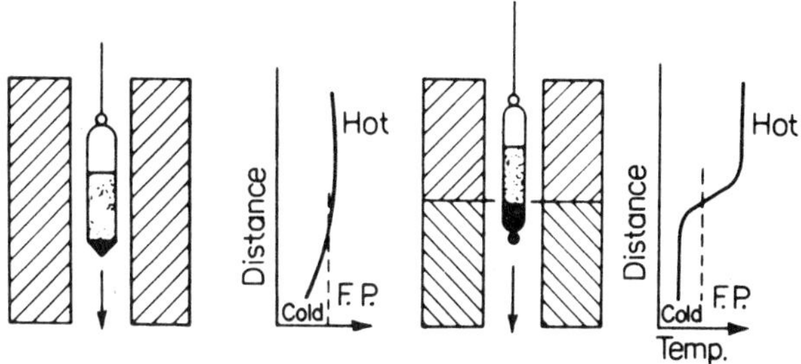

Bridgman–Stockbarger Method

Briggsian Logarithm
If $y = 10^x$, x is termed the Briggsian logarithm of y. *See also* **Brig** (**Appendix**).

Brillouin Function
In a paramagnetic substance the magnetization can be shown to be

$$M = NgJ\mu_B B_J(x)$$

where N is the number of atoms per unit volume each with a total angular momentum quantum number J and a spectroscopic splitting factor g (**Landé Splitting Factor**), μ_B is the **Bohr Magneton**, and

$$x = \frac{gJ\mu_B H}{kT}.$$

B_J is the Brillouin function

$$B_J = \frac{2J+1}{2J}\coth\frac{(2J+1)x}{2J} - \frac{1}{2J}\coth\frac{x}{2J}.$$

Brillouin Scattering
Scattering of electromagnetic waves in solids and liquids when, as a result of the scattering process, an acoustic phonon is emitted or absorbed. (An acoustic phonon is a lattice vibration in which atoms in a unit cell vibrate in the same direction. The dispersion curve for the vibrations is of the same form as for sound waves.) Brillouin scattering is analogous to **Raman Scattering**.

Brillouin Zone
If the reciprocal lattice of a crystal is obtained, but with periodicity of the lattice $2\pi/d_{hkl}$ instead of $1/d_{hkl}$, and the space be divided into

identical cells in the same way as the real lattice is divided to obtain the **Wigner–Seitz Cell**, then the cells so obtained are called Brillouin zones. These zones are very important in the theory of energy bands in crystals because, by **Bragg's Law**, the zones may be said to define a momentum or **k** space in which **k** is the wave vector for the electrons. If at the reciprocal lattice points there is strong reflection of electrons or x-rays (this will be so, unless the so-called structure factor used in x-ray analysis is zero), then there will be an energy gap at each face of the Brillouin zone. (*See also* **Ewald Sphere** *and* **Jones Zone**.)

Brinell's Hardness Test

The hardness of a material is measured by the area of indentation produced by a standard steel ball of 10 mm diameter under prescribed conditions of loading–usually 3000 kg f (29·4 kN) static load for steel and 500 kg f (4·9 kN) for softer metals. The **Brinell number** is given by

$$\frac{p}{\pi t D} = \frac{p}{\pi D\{D/2 - \sqrt{(D^2/4 - d^2/4)}\}}$$

where p is loading in kgf, t is depth of indentation, D diameter of the ball and d diameter of the indentation, all in millimetres.

Brin's Process

Formerly oxygen was obtained industrially by this process from barium oxide which was first oxidized in air to barium dioxide and then heated, when oxygen was evolved and the oxide regenerated. This method has now been supplanted by liquid-air processes.

Bromwich's Expansion Theorem (*see also* **Heaviside's Expansion Theorem**)

If
$$u = \frac{\mathrm{F}(p)}{\mathrm{f}(p)}(Gt)$$

where G is a constant, the solution is

$$u = G\left\{N_0 t + N_1 + \sum \frac{\mathrm{F}(\alpha)}{\alpha^2 \mathrm{f}'(\alpha)} e^{\alpha t}\right\}$$

where N_0 and N_1 are defined by

$$\frac{\mathrm{F}(p)}{\mathrm{f}(p)} = N_0 + N_1 p + N_2 p^2 + \ldots .$$

α is any root of $f(p) = 0$, $f'(p)$ is the first derivative of $f(p)$ with respect to p and the summation is over all the roots α. The solution reduces to $u = 0$ when $t = 0$.

Brønsted–Lowry Definition of acids and bases
According to these workers, an acid is defined as a substance with a tendency to lose a proton, and a base as a substance with a tendency to gain a proton.

Such a property of any substance cannot manifest itself unless the solvent molecules are themselves able to act as proton acceptors or donors.

Brooks–Herring Formula
An expression for the mobility of electrons in a medium containing a number of point imperfections or impurities. It takes the same form as the **Conwell–Weisskopf Equation** but replaces $1/\ln(1+x)$ by $F(3k_BT)$ where $F(E)$ is a Fermi–Dirac function. (*See also* **Fermi–Dirac Statistics.**)

Brouwer Diagram
The concentrations of defects in a binary compound can be plotted as functions of the partial pressure of one of the constituent elements. In a Brouwer diagram, concentrations and pressure are plotted logarithmically and this produces linear regions on the plots. Also called a **Kröger–Vink Diagram**.

Brownian Movement
In 1827 the botanist, R. Brown, observed under the microscope that pollen grains in water were in a continual state of haphazard motion. The erratic motion has been observed for all kinds of small particles suspended in various media. The phenomenon is explained by the continual buffeting of the suspended particles by the molecules of the medium in which they are suspended.

Brunauer, Emmett and Teller (B.E.T.) Adsorption Equation
Whereas Langmuir's equation (*see* **Langmuir Adsorption Isotherm**) assumes an adsorbed layer which is monomolecular, the B.E.T. equation assumes multimolecular adsorption with each separate layer

obeying the Langmuir equation. The adsorption equation becomes

$$\frac{V}{V_m} = \frac{cp/p_s}{(1-p/p_s)(1-p/p_s+cp/p_s)}$$

where V is the volume of gas adsorbed, V_m is the volume of gas required to cover the entire surface with a monomolecular layer, p is the pressure of the gas and p_s the saturation pressure. c contains the net heat of adsorption:

$$c = c_1 \exp\left(\frac{Q_A - Q_L}{RT}\right)$$

where Q_A is the average heat of adsorption, Q_L the heat of liquefaction of the vapour and c_1 a constant of the system.

Bucherer Reaction
A naphthol may be reversibly converted into a naphthylamine by the action of ammonia in the presence of an aqueous solution of a sulphite.

Budan's Theorem
For a real algebraic equation of the form

$$f(x) = a_0x^n + a_1x^{n-1} + \ldots a_n = 0$$

$$\text{where } a_0 \neq 0$$

if $N(x)$ is the number of changes of sign in the sequence of derivatives $f(x), f'(x), f''(x), \ldots f^n(x)$ where vanishing derivatives are disregarded, then the number of real roots located between two real numbers a and $b > a$, where a and b are not roots themselves, is either $N(a) - N(b)$ or less than $N(a) - N(b)$ by a positive even integer.

Budde Effect
The expansion observed on exposing halogen vapours to light caused by the rise in temperature resulting from heat evolved in the recombination of atoms.

Burger–Dorgelo–Orstein Rule
For small multiplet splitting in atomic spectra when **Russell–Saunders Coupling** is applicable, the sum of the intensities of all the lines of a multiplet which belong to the same initial or final state

is proportional to the statistical weight of $2J+1$ of the initial or final state respectively, where J is the quantum number for the total angular momentum of the electrons.

Burgers' Circuit
A method of determining whether an imperfection is contained in a

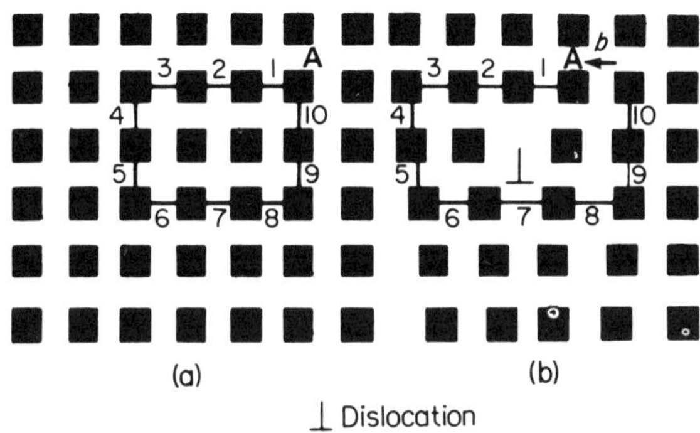

Burgers' Circuit

crystal lattice. Consider the perfect two-dimensional structure in diagram (a). The circuit marked with the arrows consists of ten moves to go from A back to A; e.g. $1 \rightarrow 2 \rightarrow 3$ and so on. In diagram (b), a dislocation is contained in the area of Burgers' circuit and it now takes more than ten moves to go from A back to A.

Burgers' Vector
The vector necessary to close **Burgers' Circuit**. It is at right angles to the dislocation for a pure edge dislocation and parallel to the dislocation for a screw dislocation.

Burnside's Formula
A formula for double integration:

$$\int_{-1}^{+1}\int_{-1}^{+1} f(x, y)\,dx\,dy = \tfrac{40}{49}\{f(a,0)+f(0,a)+f(-a,0)+f(0,-a)\} + \tfrac{9}{49}\{f(b,b)+f(b,-b)+f(-b,b) + f(-b,-b)\}$$

where $a = \sqrt{\frac{7}{15}}$, $b = \sqrt{\frac{7}{9}}$ and $f(x, y)$ is a polynomial of degree 5 at most.

Burnside's Theorem
The necessary and sufficient condition that an n-dimensional representation $D(R)$ be an irreducible representation of a group is that there exist n^2 linearly independent matrices in set $\{D(R)\}$. *See also* **Schur's Lemma**.

Burstein Effect
The variation in width of the optical energy gap of certain semiconductors with the amount of doping. Usually, the optical gap is between the highest energy state in the valence band and the lowest state in the conduction band; i.e. the smallest energy difference required for excitation of electrons from the valence to the conduction band. In some semiconductors having a very large carrier density, the first unoccupied states lie well within one of the bands and this position is a function of doping.

Buys Ballot's Law
In meteorology the law giving the direction of rotation of cyclones and anticyclones. In the northern hemisphere with wind arriving from the rear of the observer, the pressure is lower on the left-hand side than on the right; the converse is true in the southern hemisphere. *See also* **Coriolis Force**.

C

Cailletet and Mathias Law

The mean of the densities of a liquid and its saturated vapour at the same temperature is a linear function of temperature. The densities of the liquid and of saturated vapour in equilibrium with it are known as orthobaric densities and, if ρ_t is the arithmetical mean at any temperature t, then

$$\rho_t = \rho_0 + \alpha t$$

where ρ_0 and α are constant for a given substance.

Callendar's Equation

An equation representing the behaviour of superheated steam below the critical point and away from the liquid–vapour region. It is:

$$V - b = \frac{RT}{p} - \frac{a}{T^n}$$

where $n = 10/3$ and a and b are constants.

Callen Relations *See* **Heurlinger Equations**

Calzecchi–Onesti Effect

The conductivity of loosely aggregated metallic powders changes under the action of applied electromotive forces.

Cannizzaro Reaction

This reaction is an example of disproportionation. When an aldehyde having no α hydrogen atoms is heated with concentrated aqueous caustic soda, an alcohol is formed

$$2HCHO + NaOH \rightarrow H.COONa + CH_3OH$$

All aldehydes may be made to undergo this reaction in the presence of aluminium ethoxide.

Carathéodory's Principle
In the neighbourhood of an arbitrary state J of a thermally isolated system there are states J′ which are inaccessible from J. It is a statement of the second law of thermodynamics and asserts that there are states as close as we wish to J that cannot be reached by an adiabatic process. These inaccessible states have entropy less than J.

Carnot Cycle

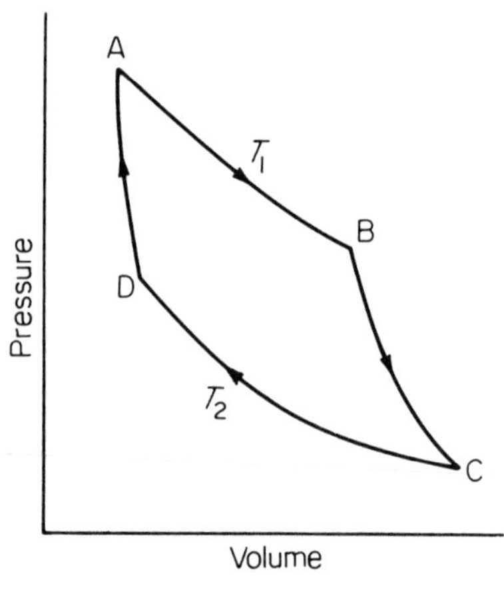

Carnot Cycle

A sequence of operations forming the working cycle of an ideal heat engine of maximum thermal efficiency. It consists of isothermal expansion AB, adiabatic expansion BC, isothermal compression CD and adiabatic compression DA as shown in the diagram. Application of the First Law of Thermodynamics leads to an equation for the efficiency of any reversible machine:

$$\frac{Q_1 - Q_2}{Q_1} = \frac{T_1 - T_2}{T_1}$$

where Q_1 is the heat absorbed at the temperature T_1 and Q_2 is that given up again at the lower temperature T_2.

Carnot's Theorem
All machines working reversibly between the same temperatures of source and sink have the same efficiency. That is to say, provided the

machine functions reversibly, its efficiency is independent of the nature of the working fluid employed or the mode of operation.

Cartesian Coordinates
A rectangular system of coordinates in which a position vector **r** can be written as the sum of its vector components along the x, y and z axes by

$$\mathbf{r} = x\mathbf{i} + y\mathbf{j} + z\mathbf{k}$$

where **i, j** and **k** are unit vectors along the x, y and z axes respectively. The system is named after Descartes, who often used the Latin version of his name, Cartesian.

Cassini's Ovals
Curves giving the locus of points having the product of their distances from $(-a, 0)$ and $(a, 0)$ as c^2 and having the form

$$(x^2 + y^2 + a^2)^2 - 4a^2x^2 = c^4$$

Castigliano's Theorems
I. If U is the strain energy (*see* **Clapeyron's Theorem**) when a body is deformed, s_i are the displacement components and F_i are the corresponding force components, then

$$s_i = \frac{\partial U}{\partial F_i}$$

II. If the prescribed values of the displacement components are denoted s_i' then the unprescribed forces F_i are such that $(U - \Sigma s_i' F_i)$ is a minimum.

Castner–Kellner Process (for the Manufacture of Sodium Hydroxide)
A chemical process of pedagogical importance involving the use of a mercury cathode in which the sodium liberated from the electrolysis of brine forms an amalgam from which the caustic soda is produced by the action of water.

Castner Process (for the Preparation of Sodium)
A process for the manufacture of sodium by the electrolysis of molten caustic soda. Due to the anodic formation of water which reacts with the sodium produced at the cathode, the method is inefficient and has

been replaced by other methods depending upon the electrolysis of brine (*see* **Downs' Process**).

Catalan's Solids
A description sometimes used for the duals of the **Archimedean Polyhedra**.

Catalan's Trisectrix
A curve having a polar equation of the form

$$r = a \sec^3 (\theta/3).$$

The curve is so-called because, with suitable construction lines, it can be used for trisecting an angle. It is also called **Tschirnhausen's Cubic** or **l'Hôpital's Cubic**.

Cauchy–Riemann Equation
A function $f(x) = u(x, y) + jv(x, y)$ is analytic if the Cauchy–Riemann equations hold:

$$\frac{\partial u}{\partial x} = \frac{\partial v}{\partial y} \quad \text{and} \quad \frac{\partial u}{\partial y} = -\frac{\partial v}{\partial x}$$

Cauchy's Boundary Conditions
Mathematical boundary conditions involving the relation between the value and the normal gradient of the required solution on the boundary.

Cauchy–Schwarz Inequality

$$\left|\sum_j a_j b_j\right|^2 \leqslant \left(\sum_j |a_j|^2\right)\left(\sum_j |b_j|^2\right)$$

Cauchy's Condensation Test
If $f(n)$ is a positive decreasing function of n, and a is any positive integer greater than 1, the two series $\Sigma f(n)$ and $\Sigma a^n f(a^n)$ are both convergent or both divergent.

Cauchy's Dispersion Formula

$$n = A + \frac{B}{\lambda^2} + \frac{C}{\lambda^4}$$

where n is the refractive index and A, B and C are constants. As A, B and C are determined from observations of n at widely different wavelengths, the formula represents the dispersion of most substances with considerable accuracy, although the theory used by Cauchy, the elastic solid theory of light, is not correct.

Cauchy's Frequency Distribution

A frequency distribution for the variable x given by

$$\mathrm{f}(x) = \frac{1}{\pi a} \frac{1}{1 + [(x - \lambda)/a]^2}$$

where the distribution is symmetric about λ, which is the median but not the mean. α is a measure of the spread of the distribution.

Cauchy's Inequality

The arithmetical mean of n positive numbers is greater than or equal to the nth root of the product of these numbers.

Cauchy's Integral Formula

If $\mathrm{f}(z)$ is analytic inside and on a closed contour C, and if a is a point, then

$$\oint \frac{\mathrm{f}(z)}{z - a} \mathrm{d}z = 2\pi \mathrm{j}\, \mathrm{f}(a).k$$

where $k = 1$ if a is inside C, $k = 1/2$ if a is on C and $k = 0$ if a is outside C.

Cauchy's Mean

Arithmetic mean: If $a_n \to a$ as $n \to \infty$, then $(a_1 + a_2 + \ldots a_n)/n \to a$. Geometric mean: if $a_n \to a > 0$ as $n \to \infty$ and $a_n > 0$, then $(a_1 a_2 \ldots a_n)^{1/n} \to a$.

Cauchy's Mean Value Theorem

If $\mathrm{f}(x)$ and $\phi(x)$ are continuous in the range $a \leqslant x \leqslant b$, and $\mathrm{f}'(x)$ and $\phi'(x)$ exist and do not vanish simultaneously in the range $a < x < b$, then there exists a point c such that $a < c < b$ where

$$\frac{\mathrm{f}(b) - \mathrm{f}(a)}{\phi(b) - \phi(a)} = \frac{\mathrm{f}'(c)}{\phi'(c)}$$

Cauchy's Rule for Series

$$\sum_{n=0}^{\infty}\left(\sum_{j=0}^{n} a_j b_{n-j}\right)=\left(\sum_{j=0}^{\infty} a_j\right)\left(\sum_{j=0}^{\infty} b_j\right)$$

provided the three series are convergent.

Cauchy's Tests for Convergence
(1) In

$$\sum_{n=1}^{\infty} u_n = u_1 + u_2 + \ldots$$

if

$$\lim_{n\to\infty} |u_n|^{1/n} < 1$$

Σu_n is absolutely convergent.
(2) If $f(x)$ be a steadily decreasing positive function such that $f(n) \geqslant a_n$, then

$$\Sigma a_n \text{ is convergent if } \int_m^{\infty} f(x)\,dx \text{ is convergent.}$$

Cauchy's Theorem (for a Square Matrix)
The adjugate of a square matrix is that matrix whose (i, j)th element is the cofactor of the (j, i)th element in $|A|$, the determinant of A; the cofactor is $(-1)^{i+j}$ multiplied by the minor, and the minor is formed from the square matrix A by suppressing the ith row and jth column. Then, Cauchy's theorem states that if A is a square matrix of order n,

$$|\text{adj } A| = |A|^{n-1}$$

Cauchy's Theorem (for the Existence of a Limit)
The necessary and sufficient condition for the existence of a limiting value of a sequence of numbers $z_1, z_2, z_3, \ldots$ is that corresponding to any given positive number ε, however small, it is possible to find a number n such that

$$|z_{n+p} - z_n| < \varepsilon$$

for all positive integer values of p.

Cavalieri's Theorem
Two bodies, having equal heights and bases, and with all sections parallel to and the same distance from these respective bases also equal, have the same volume.

Cayley–Hamilton Theorem
In matrix theory, if A is a square matrix and $|f(\lambda)| = |A - \lambda I| = 0$ is its characteristic equation where I is the unit matrix, then

$$|f(A)| = 0.$$

Cayley–Klein Parameters
Quantities used to express the orientation of a rigid body. If the parameters are represented by α, β, γ and δ, then they are related to the **Eulerian Angles** by

$$\begin{aligned}
\alpha &= \exp[j(\chi+\phi)/2]\ \cos\theta/2 \\
\beta &= j\exp[j(\chi-\phi)/2]\ \sin\theta/2 \\
\gamma &= j\exp[-j(\chi-\phi)/2]\ \sin\theta/2 \\
\delta &= \exp[-j(\chi+\phi)/2]\ \cos\theta/2
\end{aligned}$$

Cayley Numbers
These have eight base elements: $1, e_1, \ldots e_7,$

with $\quad e_i^2 = -1 \quad e_ie_j = -e_je_i$

and $\quad e_1e_2 = e_3 \quad e_1e_4 = e_5 \quad e_1e_6 = e_7 \quad e_2e_5 = e_7$

$\quad e_2e_4 = -e_6 \quad e_3e_4 = e_7 \quad e_3e_5 = e_6.$

All other products of base elements are obtained by the further rule that $e_ie_k = e_j$ implies that $e_ke_j = e_i$ and $e_je_i = e_k$.

Cayley's Sextet
A curve having a polar equation of the form

$$r = 4\,a\cos^3(\theta/3)$$

Cayley's Theorem
Every finite group is isomorphic with a certain group of permutations. Thus the study of finite groups may be reduced to a study of the permutation groups, and the theorem has important application to symmetry studies. Two groups $G(=g_1, g_2, \ldots)$ and

$F(=f_1, f_2, \ldots)$ are called isomorphic if a one-to-one correspondence can be established between their elements and if their multiplication tables coincide, i.e. if

$$g_i \leftrightarrow f_i \qquad g_j \leftrightarrow f_j \qquad g_i g_j \leftrightarrow f_i f_j$$

Celsius Scale
The original Celsius scale of temperature of 1742 had as zero the boiling point of water and as 100° the freezing point. The scale was inverted by Christin in 1743. The terms of Celsius and Centigrade are used synonymously.

Cerenkov Radiation *See* **Cherenkov Radiation**

Cesàro's Summation Formula
A method of obtaining a sum of a series Σx_i which may not necessarily be convergent. If $s_1, s_2, \ldots s_n$ are partial sums of the series, then the sum is given by

$$\lim_{n \to \infty} \frac{s_1 + s_2 + \ldots s_n}{n}.$$

For example, for the series $1 - 1 + 1 - 1 + \ldots$, the method obtains a value of 0.5.

Ceva's Theorem
If three concurrent straight lines are drawn from the vertices A, B and C of a triangle to meet the opposite sides (produced if necessary) at P, Q and R respectively, then the product $\overline{AR}\,.\,\overline{BP}\,.\,\overline{CQ}$ of three alternate segments taken in order is equal to the product $\overline{RB}\,.\,\overline{PC}\,.\,\overline{QA}$ of the other three.

Chandrasekhar Limit
No white-dwarf star can be more massive than approximately 1.4 solar masses. The limit arises because the larger the white-dwarf star, the closer it packs; the upper mass limit is set by how close the atoms can be compressed together without producing a neutron star, where the protons and electrons are forced to combine by the gravitational pressure.

Chapman Equation
The viscosity of a gas is equal to:

$$\zeta = \frac{0.499\, mv}{\sqrt{2}\pi d^2(1 + C/T)}$$

where m is the molecular mass, v its average velocity and d the collision diameter of the molecule.

C is a constant (*see also* **Sutherland's Formula**).

Chapman–Jouguet Condition
A detonation in which the velocity of the shock front with respect to the material behind it is equal to the corresponding sound velocity.

Charles' Law
At a constant pressure, the volume of a definite mass of an ideal gas is proportional to its temperature on the absolute scale. (*See also* **Boyle's Law**.)

Charlier's Checks
A method of checking the arithmetic involved in calculating the mean value x and standard deviation for a number of observations x_s of frequency f_s. It uses

$$\Sigma f_s(d_s + 1) = \Sigma f_s d_s + \Sigma f_s$$
$$\Sigma f_s(d_s + 1)^2 = \Sigma f_s d_s{}^2 + 2\Sigma f_s d_s + \Sigma f_s$$

where d_s is the deviation of x_s from a working mean m.
[Using this notation

$$x = \frac{\Sigma f_s d_s}{\Sigma f_s} + m \quad \text{and} \quad \sigma^2 = \frac{\Sigma f_s d_s{}^2}{\Sigma f_s} - (m - x)^2]$$

Charpy Test
In metallurgy, an impact test in which a notched specimen, fixed at both ends, is hit by a pendulum.

Chartier's Test
If $\mathrm{f}(x) \to 0$ steadily as $x \to \infty$ and if $\left|\int_a^x \phi(x)\,\mathrm{d}x\right|$ is bounded as $x \to \infty$, then $\int_a^\infty \mathrm{f}(x)\phi(x)\,\mathrm{d}x$ is convergent.

c

Chebyshev Polynomials
These are functions

$$T_n(x) = \frac{1}{n}\sqrt{\left(\frac{2}{\pi}\right)}\cosh\,(n \cosh^{-1} x)$$

$$U_n(x) = \frac{1}{\sqrt{(x^2-1)}}\sqrt{\left(\frac{2}{\pi}\right)}\sinh\left\{(n+1)\cosh^{-1} x\right\}$$

called Chebyshev polynomials of the first and second kind respectively. The polynomial form arises on substituting $z = \cosh x$ into these functions. They are special cases of the **Gegenbauer Polynomials** and are solutions of the differential equation

$$(x^2-1)\frac{d^2y}{dx^2} + x\frac{dy}{dx} - n^2y = 0$$

Chebyshev's Approximation
An arbitrary function can be expanded in a given finite range as a power series (e.g. **Taylor's Series**). An expansion in terms of **Chebyshev Polynomials** will ensure that the value of the deviation of the resulting polynomial from the original function will be not greater than a given pre-assigned number.

Chebyshev's Inequality
If x is a random variable of mean value $\bar{x}$ and σ is the standard deviation, then there is an upper limit to the probability that the absolute deviation of x from $\bar{x}$ is greater than or equal to a given value $a(>0)$, and this probability limit is given by

$$\mathrm{P}\left[\,|x-\bar{x}| \geqslant a\,\right] \leqslant \frac{\sigma^2}{a^2}$$

Cherenkov Radiation
Radiation which arises when a charged particle traverses a medium at a velocity greater than the phase velocity of light in that medium. The particle sets up local polarization of the dielectric medium and, as the coherent growth and decay of a sequence of dipoles occurs along the path of the particle, radiation is excited at an angle θ to the path, where $\cos\theta = c/vn$. v is the particle velocity, c is the velocity of light *in vacuo* and n is the refractive index of the medium. Thus, the direction of radiation lies in the surface of a cone of semi-angle θ, and the electric

ector of the radiation is perpendicular to the surface of this one.

Child's Law
An equation connecting the anode current I_a and voltage V_a for a diode. It is of the form

$$I_a = K V_a^\eta$$

where K and η are constants. This equation holds under space-charge-limited conditions and can be applied to vacuum and solid-state diodes.

Chladni's Figures
Patterns set up by sprinkling sand upon a glass plate which is set vibrating with a bow. The plate is supported centrally and different symmetrical patterns are obtained by gripping one edge of the plate at different points, which thereby become nodes.

Christiansen Effect
The transmission of coloured light when white light is incident on a finely powdered homogeneous and transparent substance, such as glass or quartz, which is immersed in a liquid of the same refractive index. This effect arises because, whereas with monochromatic light complete transparency can be obtained, with white light, because of differences of dispersion in the solid and liquid, refractions will only match for a narrow range of the spectrum.

Christoffel Symbols
These are three index symbols represented in the form

$$\begin{bmatrix} i \\ j\,k \end{bmatrix}$$

and are not tensors. They are used in the formation of derivatives of vectors such as divergence and curl in changing from one coordinate system to another.

Chugaev Reaction
s-methyl *o*-alkyl xanthates containing $\alpha\beta$ hydrogen to the alkyl group pyrolyse to yield olefins without rearrangement of the carbon skeleton.

Clairaut's Differential Equation
If

$$y = x\frac{dy}{dx} + f\left(\frac{dy}{dx}\right)$$

its general solution is of the form $y = cx + f(c)$.

The singular solution is obtained by eliminating dy/dx between

$$y = x\frac{dy}{dx} + f\left(\frac{dy}{dx}\right) \quad \text{and} \quad x + \frac{d}{dx}f\left(\frac{dy}{dx}\right) = 0$$

Clairaut's Theorem
If g_e and g_λ are the values of the acceleration due to gravity at the earth's equator and latitude λ respectively, and m is the ratio of the centrifugal acceleration to g_e, then

$$g_\lambda = g_e[1 + (5/2m - e)\sin^2\lambda],$$

where e = ellipticity = $(r_1 - r_2)/r_1$ and r_1 and r_2 are the equatorial and polar radii. Bouguer and Faye have given correction rules for local variation of altitude.

Claisen Condensation
A ketoester may be formed by the reaction of two esters having α-hydrogen atoms in the presence of sodium ethoxide, sodamide or triphenyl methyl sodium. In certain cases the second ester does not contain an α-hydrogen atom, viz. ethyl benzoate, ethoxide, sodamide or ethyl formate. The condensation also occurs between esters and ketones to give a 1:3 diketone. The best-known use of this reaction is the preparation of acetoacetic ester when ethyl acetate condenses with itself in the presence of sodium ethoxide.

Claisen Rearrangement
When heated to about 200° C, allyl ethers of phenols rearrange to form the corresponding *o*-allyl phenols.

Claisen–Schmidt Condensation
An aromatic aldehyde will condense with an aliphatic aldehyde or a ketone in the presence of dilute alkali to give an unsaturated aldehyde.

Clapeyron Equation of State
The combined gas equation for one gram-molecule is

$$pV = RT.$$

In order to apply this to an arbitrary quantity of gas, the number n of moles it contains must be known. The equation can then be written in the form

$$pV = nRT$$

where $R = 8{\cdot}314$ joules degree^{-1} mole^{-1}. The equation is known as the Clapeyron Equation of State for gases.

Clapeyron's Theorem
If s_i are the displacement components and F_i are the corresponding force components when a body is deformed, where $i = 1, 2, 3, \ldots$ then the strain energy is given by

$$U = \tfrac{1}{2}\Sigma F_i s_i$$

(The F_i terms include all deforming loads and body forces, but not the six constraints required to hold the body in equilibrium by preventing translation and rotation.)

Clark Process
Softening of water by the addition of lime water to convert acid carbonates to normal carbonates.

Claude Process
The liquefaction of air in stages, the expanding gas being cooled by performing work on pistons.

Claus Benzene Formula *See* **Kekulé Benzene Formula**

Clausius *See* **Appendix**

Clausius–Clapeyron Equation
A thermodynamic equation of the form

$$\frac{\mathrm{d}p}{\mathrm{d}T} = \frac{\Delta H}{T\Delta V}$$

giving the rate of change of pressure with temperature, in terms of the increase of enthalpy in a system consisting of liquid and vapour in equilibrium and of the increase in volume accompanying vaporization.

Clausius Equation
A modified gas equation of the form:

$$\left\{p+\frac{a}{T(V+c)^2}\right\}(V-b)=RT$$

where a, b and c are constants.

Clausius–Mosotti Equation
An expression for molar polarization in terms of the dielectric constant ε, the molecular weight M, and the density ρ. In SI units

$$P=\frac{\varepsilon-1}{\varepsilon+2}\cdot\frac{M}{\rho}.$$

If, in the optical range, ε is replaced by the square of the complex refractive index, n^{*2}, the **Clausius–Mosotti–Lorentz–Lorenz equation** is obtained. (*See also* **Lorentz–Lorenz Law**.)

Clausius' Statement (of the Second Law of Thermodynamics)
No process is possible whose sole result is the removal of heat from a reservoir at one temperature and the absorption of an equal quantity of heat by a reservoir at a higher temperature. (*See also* **Kelvin's Statement**.)

Clausius' Theorem
In an arbitrary heat cycle

$$\oint\frac{đQ}{T}\leqslant 0$$

where $đQ$ is an infinitesimal heat absorption. The equality applies to a reversible cycle and the inequality to an irreversible cycle.

Clausius' Virial Theorem
For a system of n particles having position vectors $\boldsymbol{r}_i (i=1, 2, \ldots n)$ and acted upon by forces $\boldsymbol{F}_i$, the average kinetic energy $\overline{T}$ of all the particles over time is

$$\overline{T}=-\tfrac{1}{2}\overline{\sum_i \boldsymbol{F}_i\cdot\boldsymbol{r}_i}$$

Clebsch–Gordan Coefficients
If D_i and D_k are irreducible representations (*see* **Schur's Lemma**) and

$$D_i \otimes D_j = \sum_k C_{ij}^k D_k$$

then C_{ij}^k are Clebsch–Gordan coefficients and the above expression, called the Clebsch–Gordan series, gives the number of times the irreducible representation D_k occurs in the **Kronecker Product** of D_i and D_j. The series permits the determination of selection rules for transitions between energy levels.

Clemmensen Reduction
A carbonyl group may be reduced to a methylene group by zinc amalgam and concentrated hydrocholoric acid. This reaction works better with ketones than with aldehydes.

Compton Effect
Important evidence for the particle or photon nature of radiation is provided by the discovery by A. H. Compton that, if monochromatic x-rays are allowed to fall on a material of low atomic weight, e.g. carbon, the scattered x-rays contain, in addition to those having the incident wavelength, others of somewhat longer wavelength. The change of wavelength is given by

$$\Delta\lambda = \frac{h}{mc}(1 - \cos\theta)$$

where θ is the angle of scattering of a photon and m is the rest mass of an electron. If E_p is the energy of the incident photon and E'_p the energy of the scattered photon, then

$$E'_p = E_p\left[1 + \frac{E_p}{mc^2}(1 - \cos\theta)\right]^{-1}$$

Compton–Getting Effect
The sidereal variation of cosmic radiation of extra-galactic origin due to rotation of the earth's galaxy.

Compton Wavelength *See* **Appendix**

Conwell–Weisskopf Equation
The scattering of electrons by singly ionized donors or acceptors in

semiconductors has been considered by these two workers. They have found that the mobility, μ, of the electron is given in SI units by

$$\mu = \frac{64\pi^{1/2}\varepsilon_r^{\,2}(2k_B T)^{3/2}}{N_e e^3 m^{*1/2}\ln(1+x^2)}$$

where

$$x = \frac{24\varepsilon_r d k_B T}{e^2}$$

N_e is the concentration of ionized donors, $2d$ is the average distance between near ionized donor neighbours, m^* is the effective mass of the electron and ε_r is the dielectric constant.

Cooper Pair
In superconductivity theory it has been shown that electrons with energy close to the **Fermi energy** can form quasi-molecular pairs called Cooper pairs which can then behave as Bose particles subject to **Bose–Einstein Statistics**.

Corbino Disc
A thin cylindrical disc perforated with a central hole. It is used for making electrical measurements in a magnetic field, with the current flowing radially and the magnetic field perpendicular to the plane of the disc. In this arrangement the **Hall Effect** is 'shorted-out' and there is a maximum magnetoresistance called **Corbino magnetoresistance**.

Coriolis Force
Owing to the earth's rotation a body moving freely upon the earth is deflected to the right in the northern hemisphere and to the left in the southern hemisphere. If the velocity of motion is v, then (when v is not too great) the component of acceleration β at right angles to the direction of motion is

$$\beta = \frac{4\pi v \sin\phi}{T}$$

where ϕ is the geographical latitude and T is the period of the earth's rotation. According to **Newton's Laws**, each acceleration may be explained by means of a force and, in this sense, we can speak of 'the deflecting force of the earth's rotation'. Attention to this force was first drawn by G. G. Coriolis and these forces are sometimes known by his name.

Cornu–Hartman Formula

A relation between the deviation D produced by a prism and the wavelength λ of the light passing through it:

$$\lambda = \lambda_0 + \frac{C}{D - D_0}$$

where λ_0, C and D_0 are constants.

Cornu's Spiral

A double spiral used in computing intensities in a Fresnel diffraction pattern, obtained by plotting

$$x = \int_0^v \cos\frac{\pi v^2}{2}\,dv \qquad \text{against} \qquad y = \int_0^v \sin\frac{\pi v^2}{2}\,dv$$

Cotes' Rule

For obtaining the area under a curve, assume for y

$$y = A_0 + A_1 x + A_2 x^2 + \ldots + A_{n-1} x^{n-1}$$

Determine the coefficients $A_1, A_2, \ldots A_{n-1}$ so that, for the n equidistant values of x, y shall have the prescribed values $y_1, y_2, \ldots y_n$. The area is then given by

$$\int y\,dx = A_0 x + \tfrac{1}{2} A_1 x^2 + \tfrac{1}{3} A_2 x^3 + \ldots + \frac{1}{n} A_{n-1} x^n$$

taken between proper limits of x.

Cotton–Mouton Effect

When light is passed through certain pure liquids in a direction which is perpendicular to an applied magnetic field, the liquids show double refraction which is much greater than would appear due to the **Voigt Effect**. It arises when the molecules of a liquid possess magnetic and optical anisotropy so that, when a magnetic field is applied, the molecules become aligned. It is proportional to the square of the magnetic field strength and is an analogous effect to the **Kerr** (electro-optic) **Effect**.

Coulomb: Coulomb, Thermal *See* **Appendix**

Coulomb Damping

The dissipation of energy of a particle in a vibrating system arising from a resistive force which is of constant magnitude (independent of both displacement and velocity) but which is always in an opposite direction to the direction of velocity of the particle.

Coulomb's Law of Force
The force between two electric charges (or magnetic poles) is proportional to their magnitudes and inversely proportional to the square of the distance between them.

Coulomb's Law of Friction
The force required just to move an object along a plane is given by μN, where N is the force normal to the plane and μ is the coefficient of friction, which is independent of the area of the surfaces in contact.

Crafts' Rule
A rule based on the **Clausius–Clapeyron Equation** for the correction of boiling points of liquids to atmospheric pressure. If T°C is the observed boiling point at p mm Hg pressure (torr; 133.322 Pa), then ΔT, the temperature difference to be added to give the boiling point at 760 mm Hg, is

$$\Delta T = c(273 + T)(760 - p)$$

where $c = 0.00012$ for most liquids.

Craik–O'Brien Illusion *See* **Mach Effect**

Cramer's Rule
If, in the equations

$$\begin{array}{l} a_{11}x_1 + \ldots + a_{1n}x_n = k_1 \\ \cdots\cdots\cdots\cdots\cdots \\ a_{n1}x_1 + \ldots + a_{nn}x_n = k_n \end{array}$$

the determinant

$$\Delta = \begin{vmatrix} a_{11} & \ldots & a_{1n} \\ \cdot & \cdots & \cdot \\ a_{n1} & \ldots & a_{nn} \end{vmatrix} \neq 0$$

then the solution of the equations is

$$x_1 = \frac{\begin{vmatrix} k_1 a_{12} & \ldots & a_{1n} \\ \cdots & \cdots & \cdots \\ k_n a_{n2} & \ldots & a_{nn} \end{vmatrix}}{\Delta} \ldots x_n = \frac{\begin{vmatrix} a_{11} a_{12} & \ldots & a_{1,n-1} k_1 \\ \cdots & \cdots & \cdots \\ a_{n1} a_{n2} & \ldots & a_{n,n-1} k_n \end{vmatrix}}{\Delta}$$

Criegee Reaction
The bond between two carboxylated carbon atoms can easily be split by oxidation. If permanganate or dichromate is used, two molecules of acid are produced. In anhydrous solution, lead tetra-acetate can be used to stop the reaction at the aldehyde stage; this reaction is known as the Criegee reaction. Periodic acid is the preferred reagent in aqueous solution.

Crookes Dark Space
As the gas pressure in a discharge tube is gradually diminished, it is found that the glow surrounding the cathode detaches itself, leaving a space around the electrode. This gap is known as the Crookes dark space. At sufficiently low pressures it fills the whole tube. The Crookes dark space is also known as the **Hittorf Dark Space**.

The Crookes dark space often is taken to include the faint red or orange glow and the **Aston Dark Space** which may occur very close to the cathode, and it is usually only the layer immediately adjacent to the cathode which is totally dark. (*See also* **Faraday Dark Space**.)

Crum–Brown and Gibson Rule
If a compound HX can be directly oxidized it is *meta*-orientating when substituted in the benzene ring, whereas if it cannot be oxidized it will be *ortho–para*-orientating. Thus, for example, the nitro group is *meta*-directing as nitrous acid is directly oxidized to nitric acid, whereas the hydroxyl group is *ortho–para*-directing as water cannot be directly oxidized to hydrogen peroxide.

Crum–Brown and Walker Synthesis
This is a general method of preparation of the even members of the series $HOOC(CH_2)_nCOOH$. If the potassium alkyl esters of dicarboxylic acids are electrolysed, the following reaction occurs

$$2(C_2H_5OOC(CH_2)_nCOO)' \rightarrow \begin{array}{c} C_2H_5OOC(CH_2)_n \\ | \\ C_2H_5OOC(CH_2)_n \end{array} + 2CO_2 + 2e$$

Curie *See* **Appendix**

Curie Point (Temperature)
Ferromagnetic materials lose their permanent or spontaneous mag-

netism when heated above a critical temperature called the Curie point. The value of this critical temperature depends upon the material. Ferroelectric materials likewise lose their spontaneous polarization above an analogous critical temperature. This upper Curie point must be clearly distinguished from the lower Curie point, below which temperature the property of ferroelectricity itself disappears.

Curie Principle
In the thermodynamics of irreversible processes, the existence of spatial symmetry properties in a system may simplify the form of the phenomenological equations in such a way that the Cartesian components of the fluxes do not depend on all the Cartesian components of the thermodynamic forces. (*See also* **Onsager's Reciprocal Relations**.)

Curie's Law
In a paramagnetic substance

$$M = \frac{C}{T}H$$

where M is the intensity of magnetization, H is the magnetic field strength and C is a constant, independent of the absolute temperature T, called the Curie constant (*see* **Curie–Weiss Law**).

Curie's Principle
The symmetry group of a medium is a subgroup of the symmetry groups of all the phenomena which may possibly occur in that medium. An alternative statement is that any symmetry in the causes will lead to symmetry of the effects. It is difficult to apply the principle to certain applications, for instance to certain cases in hydrodynamics. (*See also* **Neumann's Principle**.)

Curie–Weiss Law
In a ferromagnetic substance the magnetic susceptibility above the **Curie Point (Temperature)** is given by

$$\chi_m = \frac{C}{T - T_C}$$

where C is the Curie constant $= N\mu^2/3k_B$ and T_C is the Curie temperature. N is the number of atoms per unit volume having magnetic moment μ. The law can also be applied to electric susceptibility in ferroelectrics.

Curtius Rearrangement
Acyl azides can be prepared by the reaction of an acyl chloride on sodium azide. These acyl azides undergo rearrangement when heated to produce an isocyanate (*see* **Hofmann Rearrangement**).

Czochralski Method
Method of single-crystal growth in which a seed crystal (or alternatively, if this is not available, a wire or small capillary tube) is dipped into a melt of the material and then very slowly withdrawn. By keeping the seed crystal below the temperature of the melt, more material solidifies on to it and a large single crystal is gradually grown.

D

D'Alembertian
A differential operator

$$\square^2 V = \frac{\partial^2 V}{\partial x^2} + \frac{\partial^2 V}{\partial y^2} + \frac{\partial^2 V}{\partial z^2} - \frac{1}{c^2}\frac{\partial^2 V}{\partial t^2}$$

Thus $\square^2 V = 0$ is the wave equation for waves travelling with a velocity c.

D'Alembert's Differential Equation *See* **Lagrange's Differential Equation**

D'Alembert's Paradox
A body of any shape, completely immersed in an incompressible non-viscous fluid which is steadily moving, experiences no drag.

D'Alembert's Principle
For a system of particles of mass m_i acted upon by the real forces F_i subject to constraints such that, at a certain instant, they have acceleration a_i, then

$$F_i - ma_i = 0 \quad \text{for all} \quad i.$$

ma_i is sometimes called the 'effective force' acting on the particle, in which case D'Alembert's Principle can be stated as: the reversed effective forces and the real forces together give statical equilibrium.

D'Alembert's Test for Convergence

In $$\Sigma u_n = u_1 + u_2 + \ldots$$

if for all values of n greater than r

$$\left|\frac{u_n + 1}{u_n}\right| < \rho$$

where ρ is a positive number less than unity and independent of n, Σu_n is absolutely convergent.

Dalitz Pair
An electron–positron pair as can be produced (together with a gamma ray) by the decay of a neutral π-meson.

Dalitz Plot
In nuclear physics, a graphical method of obtaining information on the spin and intrinsic parity of a charged particle decaying into three particles.

Dalton *See* **Appendix**

Dalton's Law of Evaporation
Although the pressure of a gas over a liquid decreases the rate of evaporation, the partial pressure of the vapour at equilibrium is independent of the presence of other gases and vapours.

Dalton's Law of Partial Pressures
The total pressure of a mixture of gases is equal to the sum of the partial pressures of the constituent gases. The partial pressure is defined as the pressure each gas would exert if it alone occupied the volume of the mixture at the same temperature.

Dalton's Law of Solubility of Gases
The individual constituents of a gaseous mixture are dissolved in the ratio of their partial pressures, that is, independently of one another.

Dalton's Temperature Scale
In the Kelvin scale the unit of temperature measurement is always the same and is given by

$$t = T - 273{\cdot}16$$

where t is the temperature on the Celsius scale and T is the temperature on the Kelvin scale.

On the Dalton scale the degree is defined by the relation

$$t = 273\left\{\left(\frac{373}{273}\right)^{\tau/100} - 1\right\}$$

where τ is the temperature on the Dalton scale. Thus absolute zero on the Kelvin scale is at $-273{\cdot}16°C$ and on the Dalton scale it is at $-\infty °C$.

Darzen's Glycidic Ester Condensation

This is a reaction involving the condensation of an aldehyde or ketone with an α-haloester in the presence of sodium ethoxide, sodamide or potassium *t*-butoxide to give an α-β-epoxy ester. These glycidic esters are of interest for, on hydrolysis and decarboxylation, an aldehyde or ketone is produced with a higher carbon content than the original aldehyde or ketone used. Chloracetic ester yields an aldehyde whilst substituted chloracetic esters give rise to ketones.

Darzen's Procedure

Alkyl halides may be prepared by refluxing one molecule of an alcohol with one molecule of thionyl chloride in the presence of one molecule of pyridine.

Dauphiné Twin

A type of twinning which occurs in crystals of quartz. The twin plane is the (512) plane (*see* **Miller Indices**).

Deacon's Process

This process produces chlorine very cheaply. A mixture of air and hydrochloric acid gas is passed over a catalyst, usually cupric chloride, at 400–450°C. The yield is about 60 per cent and the excess hydrochloric acid is dissolved out of the gas. The chlorine so produced contains a large proportion of nitrogen and is used mainly for the preparation of bleaching powder.

De Broglie's Theory

The dualism observed between the wave and particle functions of radiation led De Broglie to postulate that a similar dualism might exist in material particles such as electrons. It can be shown that the wavelength λ of the hypothetical matter waves can be represented by the equation

$$\lambda = \frac{h}{mv} = \frac{h}{p}$$

where v is the group velocity of the matter waves, m is the mass of the particle and thus mv is equal to the momentum p.

De Brun–van Eckstein Rearrangement
If, for example, glucose is mixed with aqueous calcium hydroxide and allowed to stand, it isomerizes to produce a mixture of glucose, fructose, mannose, unfermented ketoses and small proportions of other products. Other aldoses and ketoses behave similarly. This rearrangement has been used to prepare certain ketoses.

Debye *See* **Appendix**

Debye Equation (for Heat Capacity of a Solid)
Debye, from a quantum-mechanical calculation of the energy of a coupled system of $3N$ oscillators, obtained an equation for the variation of the heat capacity C_V of a solid, with temperature, of the form

$$C_V = \frac{9R}{{\nu_m}^3} \int_0^{\nu_m} \frac{\exp h\nu/k_B T}{(\exp h\nu/k_B T - 1)^2} \left(\frac{h\nu}{k_B T}\right)^2 \nu^2 \, d\nu$$

where ν is the oscillator frequency and ν_m is the cut-off frequency, often called the **Debye frequency**. The quantity $h\nu/k_B$ has the dimensions of temperature and the special value $h\nu_m/k_B$, where ν_m represents the maximum frequency of the $3N$ vibration frequencies for each substance, is called its characteristic or **Debye temperature** θ. Also $h\nu/k_B T$ is dimensionless and may be replaced by the variable x. Then

$$C_V = 3R \left(\frac{12T^3}{\theta^3} \int_0^{\theta/T} \frac{x^3 \, dx}{\exp x - 1} - \frac{3\theta/T}{\exp \theta/T - 1} \right).$$

A method has been devised for evaluating the integral and the results have been tabulated for various values of θ/T. In order to determine the heat capacity of an element at any temperature, it is only necessary to know the value at one temperature so that θ may be determined; then by use of these tables the value of C_V at any other temperature may be calculated. Good agreement with experimental results has been obtained.

Debye Equation for Polarization
The total polarization **P** for N molecules is given in SI units by

$$\mathbf{P} = N\left(\alpha + \frac{p^2}{3k_B T}\right)\mathbf{E}$$

where p is the permanent electric dipole moment of the molecule, α is its polarizability and **E** is the applied electric field. The first term, due to induced dipoles in the applied electric field, will be present for all molecules; the second term arises in polar molecules having permanent dipoles.

Debye Frequency *See* **Debye Equation**

Debye–Hückel Equation
Until 1923 the activity coefficient was regarded as an empirical quantity. By taking into account the interionic attraction, Debye and Hückel found it possible to calculate the ratio of activity to concentration of an ion in dilute solution. The equation reduces, for a given solvent at a definite temperature, to:

$$-\log f_i = A z_i^2 \sqrt{I}$$

where f_i is the activity coefficient of the ion, z_i is the valency of the ion, I is the ionic strength ($\frac{1}{2}C_i z_i^2$ where C_i is the number of gram-ions per litre) and

$$A = \frac{\sqrt{2}N^2 e^3 \sqrt{\pi/1000}}{2{\cdot}303 R^{3/2}(\varepsilon_r T)^{3/2}}$$

ε_r is the dielectric constant of the solvent.

Debye–Hückel–Onsager Equation *See* **Onsager Conductivity Equation**

Debye–Hückel Screening
If a test charge is immersed in an electron gas, then the electron concentration around the test charge will be perturbed in such a way as to tend to cancel or screen the effect of the test charge. However, close to the charge and within the so-called screening length, the screening will be ineffective. Where the electron concentration is sufficiently low to use **Maxwell–Boltzmann Statistics**, the screening

is called Debye–Hückel screening and the screening length is given by

$$\left\{(4\pi e^2 n_0)/k_B T\right\}^{1/2}$$

where n_0 is the local average density of electrons.

Debye–Jauncey Scattering
This is the description sometimes given to the incoherent background scatter occurring when x-rays are scattered from a crystal. The intensity is low compared with the Bragg reflections (*see* **Bragg's Law**) at low temperatures, but increases with rise in temperature.

Debye–Scherrer Method (Powder Method)
An x-ray diffraction method in which the sample is used in the form of a powder. The sample, contained in a thin-walled silica tube or stuck to a thin fibre, is rotated in a beam of monochromatic x-radiation. The diffraction pattern is usually recorded on a cylindrical film whose axis is parallel to the axis of the specimen. The pattern consists of a series of arcs arising from the intersection of coaxial cones of radiation with the film. The geometry obeys **Bragg's Law** and the spacing of the arcs can be used to obtain the lattice parameters of the sample.

Debye–Sears Effect
A piezoelectric crystal vibrating in a liquid sets up a train of acoustic waves whose nodes occur every half wavelength. It is thus possible to obtain the effect of a diffraction grating by passing a beam of light through the liquid in a direction at right angles to the wave direction.

Debye Temperature *See* **Debye Equation**

Debye T^3 Law
At low temperatures the heat capacity of a crystalline solid is approximately proportional to T^3. This can be shown from the **Debye Equation**.

Debye–Waller Factor
When x-rays are diffracted from a crystal lattice, the intensity is reduced – due to thermal motion of the atoms – by the Debye–Waller factor

$$\exp\left(-1/3 \langle u^2 \rangle \langle \Delta k^2 \rangle\right)$$

where $\langle u^2 \rangle$ is the mean square displacement of the atoms from their equilibrium positions and Δk is the change in wave-vector of the x-rays on reflection.

Dedekind–Peirce Theorem
A class S is infinite if, and only if, it can be put into one-to-one correspondence with a proper subset of itself.

Dedekind's Definition
Suppose a certain rule divides a whole system of rationals into two classes, a lower class A and an upper class A′, so that any number in A is less than any number of A′. Either

(1) A has a greatest number or A′ has a least number, in which case the classification defines an irrational number which is to follow all in A or the least in A′, *or*
(2) if there is no greatest number in A and no least number in A′, the classification defines an irrational number which is to follow all the numbers in A and to precede all in A′.

Dedekind's Test
If Σa_n has bounded partial sums, $\Sigma|b_n - b_{n+1}|$ is convergent and $b_n \to 0$, then $\Sigma a_n b_n$ is convergent.

De Gua's Rule
If a group of r consecutive terms is missing from a polynomial $f(x) = 0$, then

(1) if r is even, the equation has at least r imaginary roots;
(2) if r is odd, there are at least $r + 1$ or at least $r - 1$ imaginary roots according as the terms which immediately precede and follow the group have like or unlike signs.

De Haas–van Alphen Effect
Oscillations of the diamagnetic susceptibility of a metal as a function of a strong magnetic field. The effect arises from the quantization of the energy of the electrons and hence the quantization of the electron orbits in the presence of a magnetic field. The latter must be sufficiently strong for electrons to complete several orbits before colliding with imperfections on the surface of the specimen.

Delambre's Analogies (or **Gauss' Analogies**)
In the solution of spherical triangles, if a, b and c are the sides of the triangle and α, β and γ are the angles opposite a, b and c respectively, then

$$\sin\tfrac{1}{2}\alpha\cos\tfrac{1}{2}(b+c) = \cos\tfrac{1}{2}a\cos\tfrac{1}{2}(\beta+\gamma)$$
$$\cos\tfrac{1}{2}\alpha\sin\tfrac{1}{2}(b-c) = \sin\tfrac{1}{2}a\sin\tfrac{1}{2}(\beta-\gamma)$$
$$\sin\tfrac{1}{2}\alpha\sin\tfrac{1}{2}(b+c) = \sin\tfrac{1}{2}a\cos\tfrac{1}{2}(\beta-\gamma)$$
$$\cos\tfrac{1}{2}\alpha\cos\tfrac{1}{2}(b-c) = \cos\tfrac{1}{2}a\sin\tfrac{1}{2}(\beta+\gamma)$$

Also **Napier's Analogies**:

$$\tan\tfrac{1}{2}(b+c)\cos\tfrac{1}{2}(\beta+\gamma) = \tan\tfrac{1}{2}a\cos\tfrac{1}{2}(\beta-\gamma)$$
$$\tan\tfrac{1}{2}(\beta+\gamma)\cos\tfrac{1}{2}(b+c) = \cot\tfrac{1}{2}\alpha\cos\tfrac{1}{2}(b-c)$$
$$\tan\tfrac{1}{2}(b-c)\sin\tfrac{1}{2}(\beta+\gamma) = \tan\tfrac{1}{2}a\sin\tfrac{1}{2}(\beta-\gamma)$$
$$\tan\tfrac{1}{2}(\beta-\gamma)\sin\tfrac{1}{2}(b+c) = \cot\tfrac{1}{2}\alpha\sin\tfrac{1}{2}(b-c)$$

See also **Napier's Rules**.

Delbrück Scattering
Scattering of photons by the coulombic field of a nucleus is called Delbrück scattering.

Delépine Reaction
Reaction of hexamethylene tetramine with primary alkyl halides yields a salt. This compound can be easily hydrolysed to give formaldehyde, ammonia and a primary amine.

Dellinger Fadeout
A failure of short-wave radio communication caused by a highly absorbing D layer in the ionosphere associated with a sunspot.

Dember Effect
The establishment of a voltage in a conductor or semiconductor by illumination of one surface. Hole–electron pairs are produced by the illumination and because, in general, the electrons are more mobile than the holes, a negative charge is established on the non-illuminated side. Hence an electric field is set up, preventing further flow. It is also called the photo-diffusion effect.

Dem'janov Rearrangement
When, say, cyclopropylmethylamine or cyclobutylamine reacts with nitrous acid, ring expansion occurs in the first case and ring contraction in the second to give the same mixture of cyclopropylcarbinol and cyclobutanol (along with a small amount of allylcarbinol). Changes in ring size due to the reaction of amines with nitrous acid are known as Dem'janov rearrangements.

De Moivre's Theorem
If n is a positive or negative integer or a fraction

$$(\cos x + \mathrm{j} \sin x)^n = \cos nx + \mathrm{j} \sin nx$$

De Morgan's Rules
These apply to **Boolean Algebra**. Any logical or binary expression equals the complement (or negation) of the expression obtained by changing all logical products (sometimes called AND statements) to logical sums (sometimes called OR statements) or vice versa and replacing variables with their complements, i.e.

$$\overline{\mathrm{A}+\mathrm{B}} = \overline{\mathrm{A}}.\overline{\mathrm{B}} \text{ and } \overline{\mathrm{A}.\mathrm{B}} = \overline{\mathrm{A}}+\overline{\mathrm{B}}$$

where $\overline{\mathrm{A}}$ and $\overline{\mathrm{B}}$ are the complements of A and B respectively.

De Morgan's Test
A series Σu_n in which $\lim_{n\to\infty} \left|\frac{u_{n+1}}{u_n}\right| = 1$ will be absolutely convergent if a positive number c exists such that $\lim_{n\to\infty} \left\{\left|\frac{u_{n+1}}{u_n}\right| - 1\right\} = -1 - c.$

Descartes' Folium
A curve having the equation

$$x^3 + y^3 = 3axy$$

Descartes' Laws
When a beam of light passes between two isotropic media, the incident and refracted rays and the normal to the surface lie in the same plane. The incident and refracted rays lie on opposite sides of the surface, and the sines of their angles of inclination bear a constant ratio depending on the media used.

Descartes' Ovals
Curves defined by the linear relationship

$$mr \pm nr' = k$$

where r and r' are distances from two fixed points. They occur as conjugate pairs as indicated by the $\pm$ sign.

Descartes' Rule of Signs
No equation can have more positive roots than it has changes of sign from positive to negative and from negative to positive, in the terms of the equation $f(x) = 0$. No equation can have more negative roots than there are changes of sign in $f(-x)$.

Destriau Effect
Certain phosphorescent inorganic materials, when suspended in a dielectric film, may be excited to luminescence if subjected to the action of an alternating electric field. This effect was discovered by G. Destriau in 1947 and such phosphors are termed electroluminescent. Another form of electroluminescence of great technological importance is carrier injection electroluminescence which occurs when a d.c. field is applied to a semiconductor. It arises because of radiative recombination of the injected carriers.

Dewar Benzene Formula *See* **Kekulé Benzene Formula**

d'Huilier's Equation
In the solution of spherical triangles

$$\tan\left(\frac{\alpha}{2} - \frac{\varepsilon}{4}\right) = +\sqrt{\left[\frac{\tan\frac{1}{2}(s-b)\tan\frac{1}{2}(s-c)}{\tan\frac{1}{2}s\tan\frac{1}{2}(s-a)}\right]}$$

where ε is the spherical excess $(= \alpha + \beta + \gamma - \pi)$ and $s = \frac{1}{2}(a+b+c)$. *See* **Delambre's Analogies** for the remaining notation.

Dido's Problem
To enclose the largest possible area within a fence of given length. It has the trivial solution of a circular fence, but the problem may be generalized with a weighting function.

Dieckmann Reaction
Certain cyclic ketones may be prepared by the Dieckmann reaction,

an intramolecular **Claisen Condensation**. Thus adipic, pimelic and suberic esters when treated with sodium give rise to five-, six- or seven-membered cyclic ketones.

Diels–Alder Reaction

A conjugated diene can be added to an ethylenic compound which has a carbonyl group adjacent to the double bond. The adduct is always a six-membered ring, the reaction taking place in the 1:4 positions. Dienes may also be made to add on to quinones, to dienes with a carbonyl group adjacent to the double bond and even, in certain cases, to the 1 : 2 double bond of another diene.

Diesel Cycle

Diesel was dissatisfied with the low efficiency of the Otto engine and was led to investigate certain contrivances whereby the efficiency of the **Carnot Cycle** may be reached. This investigation failed but led to the invention of the Diesel engine, which depends upon the idealized Diesel cycle in the diagram.

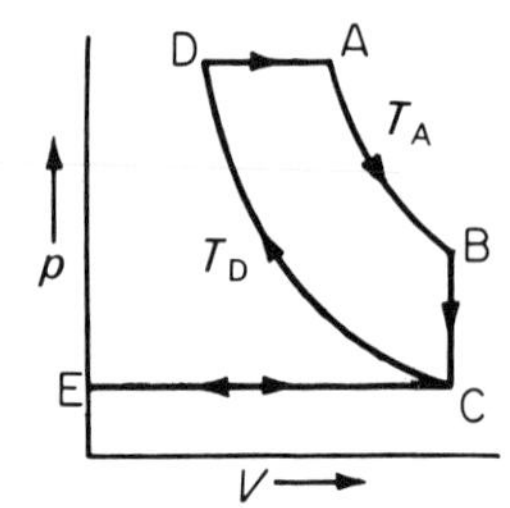

Pure air is sucked in isobarically (E → C) and compressed adiabatically (C → D). A valve is opened and oil is forced in under pressure and burns at the temperature of the system. The supply of oil is regulated so that the piston moves forward and the pressure remains constant (D → A) the temperature changing from $T_D \rightarrow T_A$. The oil is cut off at A and the gas undergoes adiabatic expansion (A → B). The valve is opened at B and the pressure drops to C. C → E is a scavenging stroke in which useless gas is forced out and the engine becomes ready for a fresh cycle.

Dieterici Equation

The Dieterici equation is a relation between the pressure of a real gas and its volume, of the form

$$p = \frac{RT}{V-b} \exp\left\{-\frac{a}{RTV}\right\}$$

where a and b are both constants. At low pressures this equation reduces to the **Van der Waals Equation**.

Diocles, Cissoid of
A curve whose equation is

$$y^2(a-x) = x^3$$

or alternatively

$$r = a\left(\frac{1}{\cos\phi} - \cos\phi\right)$$

Diophantine Equations
Equations for which solutions are required in terms of integers, but for which there is no single set of solutions.

Dirac Delta Function
A function $\delta(x)$ defined as zero when $x \neq 0$ and infinite at $x = 0$, such that

$$\int_{-\infty}^{\infty} \delta(x)\,\mathrm{d}x = 1.$$

Dirac Equation
A relativistic wave equation for an electron in an electromagnetic field. The Dirac wave-function has four components and the solution requires that, in order that the total angular momentum of the electron be constant, the electron must possess its own intrinsic angular momentum called 'spin'. The existence of the electron spin was deduced empirically by Uhlenbeck and Goudsmit before Dirac's theory was first published.

Dirichlet Condition
In the Fourier expansion of $\mathrm{f}(x)$, the assumption that, when x approaches the ends of the interval 0 and 2π, the function $\mathrm{f}(x)$ tends to definite limits which are denoted by $\mathrm{f}(+0)$ and $\mathrm{f}(2\pi - 0)$.

Dirichlet Problem
In potential theory, the determination of a function, harmonic in a region, when its boundary values are given. (*See* **Neumann Problem**.)

Dirichlet's Test for Convergence
If Σu_n converges or oscillates between finite limits and a_1, a_2,

$a_3, \ldots$ is a decreasing sequence of positive terms which tends to zero as a limit, then

$$\Sigma a_n u_n \quad \text{is convergent.}$$

Dirichlet Theorem
Every arithmetic progression a, $a+b$, $a+2b, \ldots,$ in which a, b are integers with no common divisor greater than unity, contains an infinity of primes.

Doebner–Miller Synthesis
Methylquinoline can be synthesized by this process which is analogous to the **Skraup Synthesis**. It consists of heating aniline with paraldehyde in the presence of hydrochloric acid. The mechanism probably involves the formation of crotonaldehyde ($CH_3CH{=}CHCHO$), 1:4 addition to aniline, cyclization and oxidation.

Donnan Potential
A solution of an electrolyte consisting of two diffusible ions is separated, by a membrane, from a salt consisting of the same diffusible ion as one of those on the other side of the membrane and a non-diffusible ion. At equilibrium the concentrations of the common ion are unequal on either side of the membrane. This inequality of concentration is maintained across the membrane by a potential known as the Donnan membrane potential. It can be shown thermodynamically that the fraction of the diffusible ion F_{D} which diffuses across the membrane, from the solution containing the two diffusible ions initially at concentration C_1 into the solution containing the non-diffusible ion initially at concentration C_2, is

$$F_{\mathrm{D}} = \frac{C_1}{C_1 + 2C_2}$$

It is smaller the larger the concentration of non-diffusible ion.

Donnay–Harker Principle
Generalization of **Bravais' Law** for crystal habit by considering the space group of a crystal rather than just the **Bravais Lattice** type, i.e. an extended lattice generalized from point group symmetry. When no glide planes or screw axes are present, the principle is equivalent to Bravais' law.

Doppler Effect

The observed frequency of radiation from a source possessing a component of motion in the direction of an observer differs from the true value by an amount proportional to the component. If the velocity of the source is towards the observer, the frequency appears raised; if away from the observer, the frequency is lowered. The frequency as recorded by the observer is

$$\nu' = \nu \frac{\sqrt{(1 - v^2/c^2)}}{1 - (v/c)\cos\theta}$$

where ν is the frequency of the source which has velocity v. θ is the angle between the direction of motion of the source and the direction of the observer relative to the source. Due to this effect, the outward movement of galaxies leads to a shift of the observed spectra towards the red end (*see* **Hubble Effect**).

The above expression must be obtained relativistically but approximates to the classical expression

$$\nu' = \frac{\nu}{1 - (v/c)\cos\theta}$$

for $v \ll c$.

Dorn Effect

When particles fall through water, a potential difference is set up. This is a form of electrokinetic effect.

Downs' Process

A fused electrolyte of sodium chloride containing potassium chloride and fluoride added to lower the melting point from 800°C to 600°C is electrolysed using a carbon anode and a ring-shaped cathode. The chlorine released at the anode is led off by means of a funnel over the anode, whilst the sodium collects in a ring-shaped inverted trough and is forced by the pressure of fused sodium chloride into a container. The efficiency is 70 per cent higher than that of the Castner cell (*see* **Castner Process**).

Draper Effect

At the commencement of the hydrogen–chlorine reaction there is an increase in volume at constant pressure whereas, according to the chemical equation, the volume should be constant. This increase in volume has been traced to a rise in temperature of the reactants due to the heat involved in the reaction.

Draper's Law *See* **Grotthus and Draper Law**

Drude's Equation
This relates the specific rotation $[\alpha]$ of polarized light passing through an optically active material to the wavelength λ of the incident light, and may be written as

$$[\alpha] = \frac{k}{\lambda^2 - {\lambda_0}^2}.$$

k is the rotation constant, ${\lambda_0}^2$ is the dispersion constant and λ is the wavelength of the absorbed light. Within and close to an absorption band the equation does not hold, but in regions of complete transparency an equation involving one or more Drude terms represents, in a very satisfactory way, the variation of angle of rotation with wavelength of the polarized light.

Drude's Theory of Conduction
Drude gave a formula for the specific resistance ρ of a substance, based on classical mechanical considerations, as

$$\frac{1}{\rho} = \frac{Ne^2}{m}\tau$$

where N is the number of free electrons per unit volume, e the charge and m the mass of an electron, and τ is the time for each free electron to move between two successive collisions with molecules. The formula leads to values of the mean free path of an electron which are far too large, and the difficulty becomes much worse for conduction at very low temperatures. Application of quantum mechanics has given a new approach to the subject.

Duane–Hunt Law
The high frequency limit for x-rays is proportional to the voltage applied to accelerate the electron beam used to bombard the metal target.

Du Bois Raymond's Test for Convergence
$\Sigma a_s b_s$ is convergent if $\Sigma(b_s - b_{s+1})$ be absolutely convergent and if Σa_s converge at least conditionally.

Duffing Equation
A differential equation describing the oscillations of a non-linear spring:

$$\frac{d^2y}{dt^2} + a\frac{dy}{dt} + y + by^3 = 0.$$

When $a > 0$ and $b > 0$ ('hard spring') the solution is stable even for an arbitrarily large initial disturbance. For $a > 0$ and $b < 0$ ('soft spring') the solution is only stable up to a critical value of the initial disturbance.

Dufour Effect *See* **Soret Effect**

Duhamels Theorem
Let **L** be a linear operator whose coefficients and derivatives do not involve the time variable t, and $\Phi(x, t)$ be the solution of the differential equation

$$\mathbf{L}\Phi + A(x)\frac{\partial^2\Phi}{\partial t^2} + \mathrm{B}(x)\frac{\partial\Phi}{\partial x} = 0 \qquad (0 < x < \mathbf{L}, t > 0)$$

having boundary conditions

$$\Phi(x, 0) = 0, \frac{\partial\Phi}{\partial t} = 0 \qquad (0 < x < \mathbf{L}, t = 0)$$

and boundary conditions of the form

$$\alpha\frac{\partial\Phi}{\partial x} + \beta\Phi = b(t) \qquad (x = 0, t > 0),$$

such that there are as many homogeneous boundary conditions for $x = 0$ and/or **L** as are required for solution.

Then if $\chi(x, t)$ is the solution for the special case of $b(t) = 1$ $(t > 0)$, the function $\Phi(x, t)$ is given by

$$\Phi(x, t) = \chi(x, t)b(0) + \int_0^t \chi(x, t-\tau)b'(\tau)\,d\tau$$

$$= \frac{\partial}{\partial t}\int_0^t \chi(x, t-\tau)b(\tau)\,d\tau \qquad (0 < x < \mathbf{L}, t > 0)$$

The theorem is used for obtaining the transient response of a dynamical system to any arbitrary force, and has particular application to electrical networks.

Duhem–Margules Equation

This equation shows the connection between the relative amounts of two constituents of a system and their partial vapour pressures p_A and p_B. It is assumed that the vapours behave ideally but no assumption is made about the ideality of the liquids. The equation has the form

$$x_A \frac{d \ln p_A}{dx_A} = x_B \frac{d \ln p_B}{dx_B}$$

where x_A and x_B are the respective mole fractions of the two constituents.

Duhring Rule

According to U. Duhring, if the boiling points of two substances A and B are T_A and T_B at pressure p, and $T_A{}'$ and $T_B{}'$ at pressure p'

$$\frac{T_A - T_A{}'}{T_B - T_B{}'} = \text{constant}$$

Dulong and Petit's Law

The specific heats of the elements are inversely proportional to their atomic weights. The atomic heats of solid elements are constant and approximately equal to 27 J K^{-1} mole^{-1}. Certain elements of low atomic weight and high melting point have, however, much lower atomic heats at ordinary temperatures. (*See* **Einstein's Equation for Specific Heat**.)

Dupin's Theorem

Given three families of mutually orthogonal surfaces in space, any two surfaces of different families intersect in a line which is a line of curvature for both the surfaces.

Dupré Equation

The work of adhesion W_{LS} at a gas–solid–liquid interface may be expressed in terms of the surface tensions of the three phases (γ_{LS}, γ_{GL} and γ_{GS}) by the equation

$$W_{LS} = \gamma_{GS} + \gamma_{GL} - \gamma_{LS}$$

E

Earnshaw's Theorem

The electric potential cannot be a maximum or a minimum at a point not occupied by charge. It follows that a small charged particle introduced into an electrostatic field cannot rest in stable equilibrium. The result follows from **Laplace's Equation**.

Eddington Limit

A limiting radius on a star when the radiation force on matter equals the gravitational force.

Edser and Butler's Bands

Dark bands, with constant frequency separation, in the spectrum of white light which has traversed a thin parallel-sided plate of a transparent material.

Eggertz's Method

A colorimetric estimation of carbon in steel, by dissolving the metal in nitric acid and comparing the colour with that produced by a similar metal of known carbon content.

Ehrenfest's Adiabatic Law

For a virtual, infinitely slow, alteration of the coupling conditions in a system, the quantum numbers do not change. The law is applicable both in **Bohr's Theory** and in quantum mechanics. The converse states that only such magnitudes as remain invariant for adiabatic changes can be quantized.

Ehrenfest's Equations

A second-order transition is characterized by discontinuous changes in the second-order derivative of **Gibbs' Function**. The change of pressure with temperature during such a transition is given by

$$\frac{\mathrm{d}p}{\mathrm{d}T}=\frac{C_p^f-C_p^i}{TV(\gamma^f-\gamma^i)}=\frac{\gamma^f-\gamma^i}{\kappa^f-\kappa^i}$$

where *f* and *i* refer to the two phases. C_p is the specific heat at constant pressure, γ is the coefficient of volume expansion and κ is the compressibility.

Ehrenfest's Theorem
If the position and momentum of a classical particle are replaced by their quantum-mechanical expectation values then the motion of a quantum-mechanical wave-packet satisfying **Schrödinger's Equation** will be equivalent to that defined by the classical equations of motion for the particle, provided any potentials acting upon the particle do not change over the dimensions of the wave-packet.

Ehrenhaft Effect
A helical-type movement of fine particles in the lines of force of a magnetic field during irradiation by light. The movement is similar to that of electrically charged particles in an electric field and, although the motion was originally thought to be due to 'magnetic charges' on the particles, it is caused by radiation pressure.

Einstein *See* **Appendix**

Einstein Coefficients
Suppose that, in an atom, two electronic states *n* and *m* exist ($m < n$) such that a spontaneous transition can occur from *n* to *m* with the emission of a photon. The probability that this transition will occur was given by Einstein the symbol A_{nm}. In the presence of an external radiation field of energy density $\sigma(\nu)$, where ν is the transition frequency, stimulated transitions may occur from *m* to *n*. Coefficients are defined for these stimulated transitions such that the total probabilities of transition are given by $A_{nm} + B_{nm}\sigma(\nu)$ and $B_{mn}\sigma(\nu)$. Einstein showed that the coefficients are related by the formulae

$$B_{mn} = \frac{g_n}{g_m} B_{nm} \quad \text{and} \quad A_{nm} = \frac{8\pi\nu^3 h}{c^3} B_{nm}$$

where g_n and g_m are the statistical weights of the two states *n* and *m* respectively.

Einstein–de Haas Effect *See* **Barnett Effect**

Einstein Effect
Redshift of spectral lines due to gravitation.

Einstein Frequency *See* **Einstein's Equation for Specific Heat**

Einstein Mass–Energy Equation
A fundamental equation of modern physics connecting the mass m of a particle with its energy E:

$$E = mc^2.$$

Einstein Relation (for Mobility)
A relation connecting the diffusion coefficient D and mobility μ of an ion (or charged particle) of the form

$$\mu k_B T = eD$$

where e is the charge on the particle.

Einstein's Equation for Specific Heat
It can be deduced from the classical atomic model that the specific heat of a solid at constant volume is connected to the gas constant R by the equation

$$C_V = 3R$$

This is not true, however, as all specific and hence atomic heats are temperature-dependent. By application of quantum-mechanical methods to this problem, Einstein arrived at the equation

$$C_V = 3R\left(\frac{h\nu}{k_B T}\right)^2 \frac{\exp h\nu/k_B T}{(\exp h\nu/k_B T - 1)^2}$$

where ν is the **Einstein Frequency**, the resonance frequency of the oscillator.

For most elements $h\nu/k_B T$ is sufficiently small for C_V to equal $3R$. Deviations from **Dulong and Petit's Law** are due to a high value of ν or to low temperatures. *See also* **Debye Equation**.

Einstein's Law
In the emission of electrons from metals by the incidence of radiation, if W is the work required to pass the electron through the surface, the maximum energy of the emitted electron is given by

$$\tfrac{1}{2}mv^2 = h\nu - W$$

where m is the mass and v the velocity of the electron. ν is the frequency of the radiation used and h is **Planck's Constant**. W is known as the 'work function' of the metal.

D

Otherwise expressed

$$V_e = hv - hv_0$$

where V_e is the magnitude of the retarding potential required to decrease the velocity of the electron to zero, and $v_0 = W/h$ is the minimum frequency which causes photoelectric emission from the given surface and is known as the 'photoelectric threshold'.

Einstein's Principle of Relativity

A theory of dynamics involving a space–time continuum and based on two major postulates:

I. Uniform motion of translation cannot be measured or detected by an observer stationed on a system of coordinates for measurements confined to that system.

II. The velocity of light in space is constant and independent of the relative velocities of the source and the observer.

These postulates apply to the Special Theory of Relativity where inertial frames move with constant velocities with respect to each other.

Einstein Summation Convention

In matrix and tensor notation, whenever a letter occurs as a suffix twice in a product, summation with respect to that suffix is to be automatically understood.

Elbs Reaction

A polynuclear hydrocarbon may be formed by the pyrolysis of a diaryl ketone containing a methyl group in an *o*-position to the carbonyl group.

Ellingham Diagram

A diagram which plots the change of **Gibbs' Free Energy** for a reaction as a function of temperature.

Elster and Geitel Effect

In the presence of a gas, a heated conductor assumes a charge either positive or negative. In a vacuum the charge is always negative.

Encke Roots

Let the roots of

$$x^2 + a_1 x + a_2 = 0$$

be x_1 and x_2 where $|x_2| > |x_1|$. These roots, with their signs changed, are known as the Encke roots of the equation.

Eötvös *See* **Appendix**

Eötvös Rule
This rule states that the rate of change of molar surface energy with temperature is the same for all liquids and is independent of temperature. In general the rule is unreliable, as the molar surface energy varies greatly for different liquids and is not independent of temperature.

Eratosthenes, Sieve of
The prime numbers not greater than N can be found by writing all numbers up to N and then removing those numbers that are multiples of 2, 3, . . . etc., continuing until all the primes not greater than $\sqrt{N}$ have been removed.

Erlenmeyer Azlactone Synthesis
The anhydrides of α-acylamino acids are formed by the condensation of an aromatic aldehyde with an aryl derivative of glycine in the presence of sodium acetate and acetic anhydride.

Esaki Effect
A tunnelling effect which occurs in heavily-doped *p–n* junctions with narrow space-charge layers. If the voltage across the junction is gradually increased, the current increases linearly to a peak value and then decreases to a broad minimum before increasing again. The peak in current arises when the voltage bias causes electrons in the conduction band on the *n*-type side to be at the same energy level as unoccupied states (holes) in the valence band on the *p*-type side. This results in tunnelling of electrons to these unoccupied states. Further voltage bias raises the bottom of the conduction band of the *n*-type material above the valence band of the *p*-type and then tunnelling ceases. The effect has application to the tunnel diode.

Eschweiler–Clarke Reaction
Formaldehyde can be used to methylate primary and secondary amines in the presence of formic acid.

Ètard Reaction
Aromatic aldehydes may be prepared by oxidizing the alkyl benzenes

with chromyl chloride in carbon tetrachloride solution. The complex which is precipitated is decomposed with water.

Ettingshausen Effect
If a current is passed down a conductor which is placed in a magnetic field that is orthogonal to the direction of current flow, then a temperature gradient is established at right angles to both the current and the magnetic field, and is given by

$$\frac{\mathrm{d}T}{\mathrm{d}y} = PB_z I_x$$

where P is the Ettingshausen coefficient, B_z the magnetic flux density and I_x the current.

Euclidean Geometry
The study of ordinary two- or three-dimensional space with certain Euclidean axioms: e.g. the shortest distance between two points is a straight line; two halves of equal objects are equal, etc. It can also be used to mean the study of Euclidean space of any number of dimensions. This space may be defined as all sets of n numbers $(x_1, x_2, \ldots x_n)$, where the distance $\rho(x, y)$ between $x = x(x_1, x_2, \ldots x_n)$ and $y = y(y_1, y_2, \ldots y_n)$ is defined as

$$\rho(x, y) = \left[\sum_{i=1}^{n} |x_i - y_i|^2\right]^{1/2}$$

The distance is real or complex according as x is a real or complex number.

Euclid's Algorithm (for the Highest Common Factor)
If X and Y are polynomials of degree n and m respectively, where $n \geqslant m \geqslant 1$, then their highest common factor can be found as follows: divide X by Y and obtain a remainder R_1, where R_1 is of degree $m_1 < m$. If $R_1 \neq 0$, divide Y by R_1 to obtain a remainder R_2 of degree $m_2 < m_1$ and divide R_1 by R_2. Continue until some $R_i = 0$, when R_{i-1} is the highest common factor, or until R_i is a constant, in which case the highest common factor is 1.

Eudoxus' Theorem *See* **Archimedes' Axiom**

Euler Diagram *See* **Venn Diagram**

Euler Force
The critical force required to cause the buckling of a beam supported at both ends. It is given by

$$\frac{\pi^2 EI}{l^2}$$

where E is **Young's Modulus** for the beam, I the moment of inertia about a horizontal axis through the centre of mass and l the length. The beam may not necessarily completely collapse.

Eulerian Angles
In considering the kinematics of solid bodies, a set of angles serving both to define the orientation of the solid body and as independent variables in the equations of motion.

If x, y and z are the three axes fixed in the body, and X, Y, Z represent a system of axes whose direction in space is fixed, θ is the angle between the Z and z axes. The line in which the xy plane cuts the XY plane is called the line of nodes and the angle between this and the x axis is termed ψ. The angle from the X axis measured in the positive direction round Z to the line of nodes is called ϕ. (Alternative definitions of the Eulerian angles are sometimes used.)

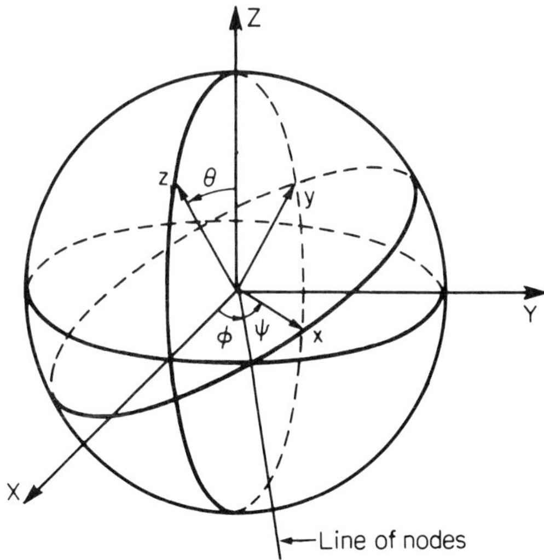

Eulerian Angles

Euler–Lagrange Equations
In variational calculus, the following problem is considered. Given a function

$$Y\left(y, \frac{\mathrm{d}y}{\mathrm{d}x}, x\right)$$

which is a function of y where y itself is a function of the independent variable x, it is required to minimize the variation of the following integral

$$\mathscr{L} = \int_a^b Y\left(y, \frac{\mathrm{d}y}{\mathrm{d}x}, x\right)\mathrm{d}x$$

with respect to the choice of the function Y. The required Y which gives $\delta\mathscr{L} = 0$ is a solution of the Euler–Lagrange differential equation

$$\frac{\mathrm{d}}{\mathrm{d}x}\left(\frac{\partial Y}{\partial \dot{y}}\right) = \frac{\partial Y}{\partial y}$$

where $\dot{y} = \mathrm{d}y/\mathrm{d}x$.

Lagrange's Equations of Motion for zero potential energy are a special case of the above.

Euler–Maclaurin Formula
If $\mathrm{f}(x)$ is known to have the $(n+1)$ values $y_0, y_1, \ldots y_n$ at $(n+1)$ points within the interval $(a, a+nh)$

$$\int_a^{a+nh} \mathrm{f}(x)\,\mathrm{d}x = h\left(\frac{y_0}{2} + y_1 + y_2 + \ldots + \frac{y_n}{2}\right) - \sum_{\text{odd } r} \frac{h^{r+1}}{(r+1)!} B_{r+1}\left(y_n^{(r)} - y_0^{(r)}\right)$$

where $y_n^{(r)}$ and $y_0^{(r)}$ are the rth derivatives at the points a and $a+nh$, and B_{r+1} is the $(r+1)$th Bernoullian number (*see* **Bernoulli Polynomials**).

Euler's Constant (or **Euler–Mascheroni Constant**)
If $u_n = 1 + 1/2 + 1/3 + \ldots + 1/n - \ln n$, then, as $n \to \infty$, $u_n \to \gamma$, where

γ is known as Euler's constant and is equal to 0·577215 . . .

$$\text{Also}\quad \gamma = -\int_0^\infty e^{-t}\ln t\,dt$$

$$= -\int_0^1 \ln\left\{\ln(1/s)\right\} ds$$

Euler's Definition of the Gamma Function

$$\Gamma(x) = \lim_{n\to\infty} \frac{n!\ n^{x-1}}{x(x+1)(x+2)\ldots(x+n-1)}$$

Alternatively, the definition

$$\Gamma(x) = \int_0^\infty e^{-t}t^{x-1}\,dt$$

for $x > 0$ and real, is called Euler's integral of the second kind.

Euler's Equation

In the motion of a perfect fluid

$$\frac{\partial \mathbf{v}}{\partial t} + \mathbf{v}\,.\,\mathrm{grad}\,\mathbf{v} = \mathbf{F} - \frac{1}{\rho}\nabla p$$

where $\mathbf{v}$ is the linear velocity of an element of the fluid at a point acted upon by a field of force $\mathbf{F}$ per unit mass of the fluid and also by the pressure p due to the remainder of the fluid. ρ is the density of the fluid.

Euler's Equations of Motion

For a rigid body fixed at one point and with coordinates fixed by the principal axes of the body, the equations of motion can be written as

$$I_{xx}\dot{\omega}_x + (I_{zz} - I_{yy})\omega_y\omega_z = L_x$$
$$I_{yy}\dot{\omega}_y + (I_{xx} - I_{zz})\omega_x\omega_z = L_y$$
$$I_{zz}\dot{\omega}_z + (I_{yy} - I_{xx})\omega_y\omega_x = L_z$$

where I_{ii} is the moment of inertia, L_j is the torque about a principal axis, ω_k is the angular momentum about the same reference line and $\dot{\omega}_k = d\omega_k/dt$.

Euler's Formula (for a Network)
A formula connecting the number of faces (F), vertices (V) and edges (E) for a map or finite-connected planar network:

$$F+V=E+1$$

Similarly, Euler's formula for a polyhedron gives:

$$F+V=E+2$$

Euler's Formula (for Curvature)
There exist, at any point of a surface, two mutually perpendicular directions lying in the tangent plane for which the curvature $1/R$ has a maximum or minimum. (R is the radius of curvature of the normal section for which the radius vector is taken as tangent.) If the curvatures corresponding to these mutually perpendicular directions are $1/R_1$ and $1/R_2$, then the curvature of any normal section is given by Euler's formula:

$$\frac{1}{R}=\frac{\cos^2\theta}{R_1}+\frac{\sin^2\theta}{R_2}$$

where θ is the angle which the tangent to the normal section forms with the direction that gives $1/R_1$.

Euler's Integrals
Euler's integral of the first kind is defined as

$$\beta(p,q)=\int_0^1 t^{p-1}(1-t)^{q-1}\,\mathrm{d}t$$

for $p>0$, $q>0$ and p and q real, where $\beta(p,q)$ is the beta function given by

$$\beta(p,q)=\frac{\Gamma(p)\,\Gamma(q)}{\Gamma(p+q)}$$

Euler's integral of the second kind is an integral expression for the gamma function. (*See also* **Euler's Definition of the Gamma Function.**)

Euler's Kinematical Theorem
Every displacement of a rigid body is equivalent to a motion of translation plus a rotation about an axis passing through a point in the body.

Euler's Numbers

If the function $1/\cos Z$ is expanded into a Maclaurin's series

$$1+\frac{E_1}{2!}Z^2+\frac{E_2}{4!}Z^4+\frac{E_3}{6!}Z^6+\ \ldots\ +\frac{E_n}{n!}Z^n+\ \ldots$$

E_n is called the nth Euler number.

Euler's Relation

$$e^{jkx}=\cos kx+j\sin kx$$

Euler's Theorem

If u is a homogeneous function of the nth degree of r variables $x_1, x_2, \ldots x_r$, i.e.

$$u(ax_1, ax_2, \ldots ax_r)\equiv a^n u(x_1, x_2, \ldots x_r)$$

and is continuously differentiable, then

$$\left(x_1\frac{\partial}{\partial x_1}+x_2\frac{\partial}{\partial x_2}+\ldots+x_r\frac{\partial}{\partial x_r}\right)^m u=n^m u$$

where m may be any integer including zero.

Euler's Transformation

$$\begin{aligned}S&=a_0+a_1x+a_2x^2+\ldots\\&=\frac{1}{1-x}a_0+\frac{1}{1-x}\sum_{k=1}^{\infty}\left(\frac{x}{1-x}\right)^k\Delta^k a_0\end{aligned}$$

where

$$\Delta^k a_0=\sum_{m=0}^{k}(-1)^m\frac{k!}{m!(k-m)!}a_{k+n-m}$$

The second series may converge more rapidly than the first.

Everett's Interpolation Formula

A modified form of the **Newton–Gauss Interpolation Formula**.

Evershed Effect

Spectrographic evidence that the general motion of gases in the penumbral regions of sunspots is radially outwards. The radial velocities are greatest near the solar limb and least at the centre of the disc.

Ewald Sphere

The distance between planes (h, k, ℓ) may be represented by $d_{hk\ell}$ in a crystal lattice. For certain purposes it is convenient to construct the reciprocal lattice (*see for example* **Brillouin Zone**). Each point in this reciprocal lattice represents a family of planes in the *direct* lattice. Any such point lies on the normal through the origin to the appropriate set of planes in the direct lattice and at a distance $d^*_{hk\ell}$ inversely proportional to the distance between these planes. Thus we may write

$$d^*_{hk\ell} = \frac{K}{d_{hk\ell}}.$$

The constant K is arbitrary, but is usually made equal to unity or 2π.

Because the reciprocal lattice represents sets of planes in the real lattice, certain of these points may give Bragg reflections for an x-ray beam of given wavelength λ and direction. It can be shown that if the direction of the x-ray beam is represented by OX in the figure and P is

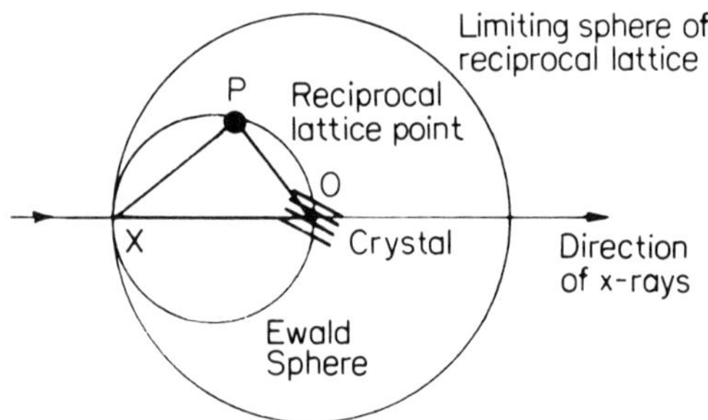

a lattice point, then the reciprocal lattice points which give Bragg reflections lie on the circle through O and X with radius $2\pi/\lambda$ when K is 2π. In three dimensions this circle becomes a sphere and is known as the Ewald sphere. If the crystal is now rotated about some axis, then diffraction can occur provided the end of the scattering vector still lies on the surface of the corresponding Ewald sphere. Thus whatever the angle of the crystal, the reciprocal lattice is bounded for diffraction by another sphere called the limiting sphere.

These concepts can simplify the interpretation of x-ray and electron diffraction photographs.

Eyring Equation
An equation in reaction kinetics connecting the rate of reaction R and the concentration of activated complexes C^*

$$R = C^* (k_B T/h) C_T$$

where the activated complexes are formed by collision of reacting molecules possessing more than a certain energy. $k_B T/h$ is the vibration frequency of a bond in classical theory and C_T is a factor which accounts for the probability that the complex will break down into the products of the reaction rather than revert to the reactants.

Eyring Formula *See* **Norris–Eyring Reverberation Formula**

F

Faber Flaws

Regions of strain in superconductors that act as nucleation centres for growth of superconducting regions.

Fabry–Pérot Fringes

Fringes formed by interference between multiple beams reflected from the inner silvered surfaces of two parallel plates. The fringes, formed at equal inclinations, consist of rings, and are similar in appearance to **Newton's Rings** for monochromatic light. The higher the reflectivity of the silvered surfaces, the sharper are the fringes. It is not practical to observe white-light fringes in this way.

Fahrenheit Scale

A scale of temperature with 32° at the freezing point and 212° at the boiling point of water at normal pressure. $32°F = 0°C$; $212°F = 100°C$.

Fajans' Precipitation Rule

A radio-element will be co-precipitated with another substance if the conditions of its precipitation are such that the element would form a sparingly soluble compound if it were present in weighable quantities.

There are many exceptions to this rule, but, as modified by Hahn, the following is strictly true:

However greatly diluted, an element will be carried down by a crystalline precipitate, provided that it can be built into the crystal lattice of the precipitate. (Also known as **Paneth's Adsorption Rule**.)

Fajans' Rules

Electrovalent bonds are most readily formed by:

I. An atom which yields an ion with a low charge, for example Na^+.

II. An atom which gives rise to a large positively charged ion or a small negatively charged ion.

However, there are exceptions to the rules.

Farad; Farad, Thermal; Faraday *See* **Appendix**

Faraday Dark Space
In a discharge tube, the dark space occurring between the positive column and the negative glow.

As the pressure is gradually reduced in a discharge tube filled with, say, air, the discharge along the tube can be separated into the following regions, from anode to cathode:

(1) the pink glow immediately around the anode;
(2) the positive column of luminous gas extending from the anode to the beginning of
(3) the Faraday dark space;
(4) the negative glow, a pale violet discharge abruptly terminating at the
(5) **Hittorf** or **Crookes Dark Space**;
(6) a red or orange glow;
(7) a very thin totally dark space adjacent to the cathode called the **Aston Dark Space**.

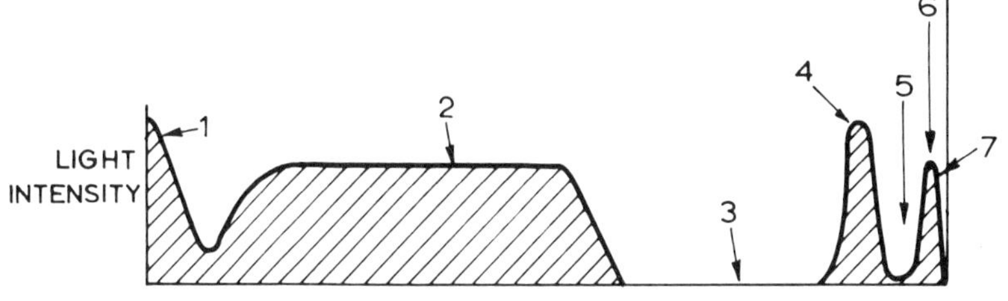

Faraday Dark Space

Faraday Effect
When a transparent isotropic medium is placed in a magnetic field, it is capable of rotating the plane of polarization of the light travelling parallel to the lines of magnetic force. (*See also* **Verdet's Constant**.)

Faraday's Laws (of Electrolysis)

I. In electrolytic decomposition the number of ions charged or discharged at an electrode is proportional to the current passed.
II. The amounts of different substances deposited or dissolved by

the same quantity of electricity are proportional to their equivalent weights.

Faraday's Laws (of Electromagnetic Induction)

I. When the flux of magnetic induction through a circuit is changing, an electromotive force is induced in the circuit.
II. The magnitude of the electromotive force is proportional to the rate of change of the flux.

Farey Sequence

All fractional (rational) numbers between 0 and 1 can be arranged in a sequence of groups such that, if each number is denoted p/q, the numbers in each group have $(p+q)$ a constant $(= n$ where $n = 1, 2, \ldots)$. Thus the sequence is

$$\frac{0}{1}; \frac{1}{1}; \frac{1}{2}; \frac{1}{3}; \frac{1}{4}, \frac{2}{3}; \frac{1}{5}; \frac{1}{6}, \frac{2}{5}, \frac{3}{4}; \frac{1}{7}, \frac{3}{5}; \frac{1}{8}, \frac{2}{7}, \frac{4}{5}; \ldots$$

wherein each group is separated by semicolons. The group having $(p+q) = n$ is called the Farey series of order n and each group is irreducible.

Favorskii Rearrangement

α-halogenated ketones in general undergo displacement of the halogen with great ease. Strong bases frequently yield acids or esters.

Fechner's Law

The eye can distinguish differences in illumination which are a constant small fraction (approximately 0.01) of the total illumination; i.e. it is the ratio rather than the arithmetic difference which is important.

Federov Parallelohedra

Polyhedra which can be placed next to each other in parallel orientation, with touching faces matching completely, so as to fill all space without gaps or overlaps. There are five typical parallelohedra: the cube; hexagonal prism; rhombic dodecahedron; elongated rhombic dodecahedron; and the cubo-octahedron. Others can be obtained by stretching or shearing these five.

Fehling's Solution
Aldehydes reduce this solution, which contains a copper complex of tartaric acid in alkaline solution, to red cuprous oxide.

Féjer's Theorem
A **Fourier Series** for the function $f(x)$ is summable for all points x at which two limits $f(x \pm 0)$ exist.

Fermat's Last Theorem
The equation $x^n + y^n = z^n$ where $x, y, z, \neq 0$ and $n > 2$ is impossible for integers x, y, z and n. This has never been proved for all values of n, although Fermat stated that he had a proof which was, however, not disclosed.

Fermat's Principle
When light passes from one point to another, the path traversed by the light ray is extremal (a minimum, maximum or some other type of 'stationary' or 'critical' point) with respect to the time required to traverse it. Fermat originally stated the principle for minimum time required and, for many practical applications, it is this case which is applicable.

Fermat's Spiral
A spiral of the form

$$r^2 = a^2\theta.$$

As there are plus and minus values of r corresponding to any positive value of θ, the curve has central symmetry about the pole.

Fermat's Theorem
If $f(x)$ is continuous in (a, b), has a derivative at every point of the interval, and attains the greatest or least value at some interior point $x = C$, the first derivative must be zero at $x = C$.

Fermat's Theorem
If p is a prime, and a any number prime to p (i.e. not divisible by p), then

$$(a^{p-1}) - 1$$

is divisible by p.

Fermi *See* **Appendix**

Fermi Constant
An interaction or coupling constant for interactions involving a total of four Fermi–Dirac particles including the decay products. It has a value of approximately 10^{-36} J m^3. Examples of Fermi–Dirac particles that may be involved are the neutron, proton, electron, neutrino and muon.

Fermi–Dirac Statistics
The study of the probability of occupation of energy states in a quantized system by indistinguishable non-interacting particles which are subject to the **Pauli Exclusion Principle**, so that there is a maximum of one particle per state. The probability that an energy state E is occupied is given by the **Fermi–Dirac distribution law**:

$$\mathrm{f}(E) = \{\exp[(E - E_{\mathrm{F}})/k_{\mathrm{B}}T] + 1\}^{-1}$$

where E_{F} is called the **Fermi energy** and is the energy at which the probability of occupation is 0.5. Fermi–Dirac statistics have particular application to electrons in metals. Particles which are subject to Fermi–Dirac statistics are called Fermi–Dirac particles or **fermions**. (Compare with **Bose–Einstein Statistics**.)

Fermi Level (Fermi Energy) *See* **Fermi–Dirac Statistics**

Fermi Plot *See* **Kurie Plot**

Fermi Selection Rules (for Beta Decay)
Fermi suggested that if the total angular momentum of a nucleus is I, then, for the emission of a beta particle, $\Delta I = 0$ (the emitted beta particle and neutrino have opposite spins) and there is no change of parity (i.e. no change of right–left symmetry). It was later found that the beta particle and the neutrino could have parallel spins, giving rise to the **Gamow–Teller Selection Rules**:

$\Delta I = \pm 1$ or 0 (units of $h/2\pi$) with no change of parity, but with $I_{\mathrm{i}} = 0$ to $I_{\mathrm{f}} = 0$ not allowed where subscript i refers to the initial momentum and f the final momentum. Beta decay is a so-called 'weak interaction' and it is now known that parity is not always conserved in weak interactions.

Fermi's 'Golden' Rule

If a quantum system has two energy levels and is subject to an external perturbation, the transition probability between the lower state and the upper state at resonance is found to be proportional to the square of the matrix element of the perturbing term and to the density of final states available: i.e.

$$\rho_{kn} = \frac{\pi |H'_{kn}|^2}{\hbar} \rho(E_k)$$

where k and n are the final and initial states, H'_{kn} is a matrix element obtained by integration over spatial coordinates and $\rho(E_k)$ is the density of final states. For resonance, the applied frequency must equal $\omega_{kn} = (E_k - E_n)\hbar$ where E_k and E_n are the energies of states k and n respectively.

Fermi Surface

In a metal the Fermi energy is a function of magnitude and direction of the momentum of the electrons. Hence the lengths of momentum vectors for electrons at the Fermi energy define a three-dimensional Fermi surface. If the energy of the electrons is proportional to the square of momentum, then the Fermi surface is a sphere. The velocity corresponding to the Fermi energy is called the **Fermi velocity** and the wavelength calculated from the wave-vector is the **Fermi wavelength**. *See also* **Fermi–Dirac Statistics**.

Fermi Temperature

The Fermi temperature of an electron gas is defined by E_F/k_B, where E_F is the Fermi energy. It is not the actual temperature of the gas, but corresponds to the temperature at which the electron gas would become classical in its statistical behaviour.

Ferranti Effect

The rise in voltage which takes place at the end of a long transmission line when the load is disconnected.

Feuerbach Circle (Nine-point Circle)

If AD, BE and CF are the altitudes of triangle ABC, H is the orthocentre and X, Y, Z, P, Q and R are the mid-points of BC, CA, AB, HA, HB and HC respectively, then the circle through X, Y and Z also passes through P, Q, R, D, E and F.

Feynman Diagram

A method used in calculations involving interactions between many bodies (for instance nuclear particles) to show how a propagating particle starting at point r_1 at time t_1 moves to r_2 at time t_2 interacting during its passage with one or more other particles in the system. The propagation is divided into the probabilities for different types of interaction using the notation:

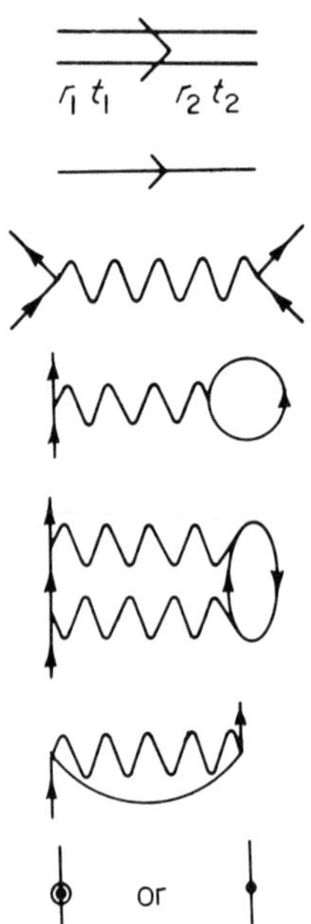

Total motion of a free particle from r_1 to r_2 in time t_1 to t_2.

Probability of free motion from r_1 to r_2.

Probability of interaction between a particle at r_1 and a particle at r_2.

Probability of instantaneous forward scattering with no change of momentum.

Probability of a chain of collision processes in which momentum at the beginning equals momentum at the end (analogous to the flow of current in an electrical network where 'flow in' equals 'flow out').

Probability of exchange scattering where the incoming particle is exchanged for the target particle.

Scattering at 0.

The important feature of a Feynman diagram is how the various lines are connected i.e. the topology of the diagram.

Fibonacci Sequence

A sequence of numbers 1, 1, 2, 3, 5, 8, 13, 21, . . ., starting from 1, such that each term is the sum of the two preceding terms. One of the properties of the series is that the ratio of successive terms becomes closer and closer to the 'Golden Section', the ratio much used in classical architecture and art. The sequence of Fibonacci numbers after the first four terms is almost indistinguishable from an exponential sequence of numbers.

Fick's Laws

I. The flux of particles J diffusing across a plane in unit time is given by

$$J = -D\frac{\mathrm{d}c}{\mathrm{d}x}$$

where D is the diffusion coefficient (the flux per unit concentration gradient) and $\mathrm{d}c/\mathrm{d}x$ is the concentration gradient.

II. The change in concentration of the diffusing particles with time ($\partial c/\partial x$) for unsteady conditions ($\partial c/\partial x \neq 0$) is given by

$$\frac{\partial c}{\partial t} = D\frac{\partial^2 c}{\partial x^2}$$

Fieser's Solution

Oxygen may be removed from nitrogen by passing it through Fieser's solution. This solution is prepared by dissolving 20 g of potassium hydroxide in 100 ml of water, adding 2 g of sodium anthraquinone β-sulphonate and 15 g of 85 per cent sodium hyposulphite (sodium thiosulphate.) The red solution is ready for use when it cools to room temperature and will absorb about 750 ml of oxygen.

Finsen Unit *See* **Appendix**

Fischer–Hepp Rearrangement

The N-nitroso derivative of a secondary amine will rearrange to give the o- and p-nitroso-amine

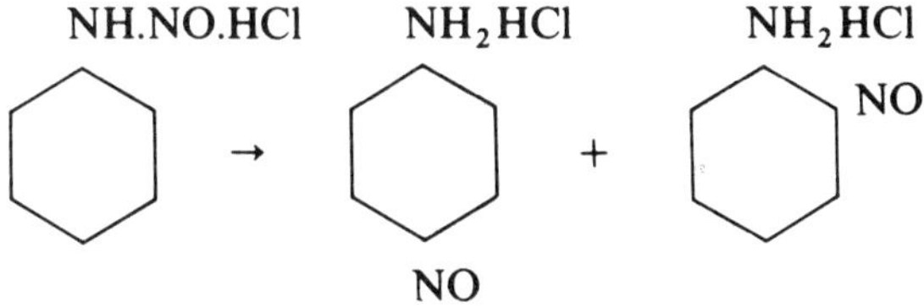

This reaction is a specific case of a more general reaction. The compound

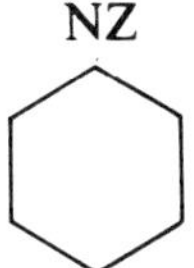

where Z is an alkyl halogen, amino, nitroso or nitro group, will also undergo rearrangement in the same way.

Fischer Indole Synthesis
This is the most important method of preparing indole derivatives. If the phenylhydrazine, or substituted phenylhydrazone of the appropriate aldehyde, ketone or ketonic acid is heated with an alkaline solution of picric acid, in the presence of zinc chloride as catalyst, ring closure is effected.

Fischer Polypeptide Synthesis
The first general method of synthesizing polypeptides in which an α-halogenated acid halide is allowed to react with an amino acid. The α-halogen atom is subsequently replaced by an amino group. The dipeptide so formed is then reacted with another α-halogenated acid chloride to give a tripeptide. Fischer was able to produce a polypeptide containing eighteen amino acid residues. It is now only of historic interest, as it cannot be used with amino acids which contain other functional groups.

Fischer–Speier Esterification
Aliphatic esters can be prepared by refluxing the respective alcohol and acid together. Such a reaction mixture may take several days to reach equilibrium. Fischer and Speier introduced the use of a mineral acid as a catalyst. Equimolar quantities of organic acid and alcohol give a poor yield, and it is usual to have the acid in considerable excess. The yield is good with primary alcohols, fairly good with secondary alcohols and poor with tertiary alcohols.

Fischer–Tropsch Gasoline Synthesis
Water gas mixed with half its volume of hydrogen heated at 200–300° C under a pressure of 1–200 atmospheres yields a mixture of hydrocarbons. The best catalyst is cobalt 100 parts, thoria 5 parts, magnesia 8 parts, kieselguhr 200 parts, but the reaction gas must be freed from sulphur which poisons the catalyst.

Fisher's *z* Distribution
If s_1 and s_2 are two independent estimates of the variance (square of the standard deviation) of a normal or Gaussian distribution (*see*

Gauss' Error Curve) then

$$z = \tfrac{1}{2}\ln (s_1/s_2)$$

and is a measure of the random sampling distribution.

Fittig Reaction

This reaction is analogous to the **Würtz Reaction**. When an aryl halide is treated with sodium in ethereal solution, the diaryl is formed.

$$2C_6H_5Br + 2Na \rightarrow C_6H_5 . C_6H_5 + 2NaBr \ (20\text{–}30\%).$$

As side-reaction products, *o*-diphenyl benzene and triphenylene are obtained.

Fitzgerald–Lorentz Contraction Hypothesis

In order to account for the null result of the Michelson–Morley experiment to determine the actual velocity of aether drift at any point, Fitzgerald in 1893 and Lorentz in 1895 suggested independently that the failure to detect velocity of motion through the aether was due to shrinkage of the apparatus used in the Michelson–Morley experiment by an amount which exactly concealed the motion.

Shortly, an arm of length l, when at rest, would, when moving with velocity u longitudinally through the aether, contract in a ratio $(1 - u^2/c^2)^{1/2}$ as a result of its motion.

Fizeau Fringes

Interference fringes observed in a wedge-shaped film due to interference between waves reflected from the two surfaces. With an extended white-light source, the fringes coincide at the film and are said to be localized in the plane of the film. With a point source the fringes are no longer localized in this way.

Flade Potential

After a metal has become passivated, the decay of passivity with time is represented firstly by a steep fall of potential in the active direction; then by a less steep change; and, finally, by a steep descent to the active value. The value of potential immediately preceding the final step is known as the 'Umschlagspunkt' or Flade potential.

Fleming's Rule

A rule relating the direction of flux, motion and e.m.f. in an electric machine. The forefinger, second finger and thumb, placed at right angles to each other, represent respectively the direction of flux, e.m.f. and torque. If the right hand is used, the conditions are as for a generator, whereas the left hand gives the relation between these quantities for a motor.

Flood's Equation
In a binary fused salt system, say, $M_1A_1 + M_2A_2$, the liquidus temperature is given by

$$T_l = T_m\left(\frac{1 + X^2\Delta G/L}{1 - [2RT_m \ln(1 - X)]/L}\right)$$

where T_m is the melting point (in kelvins) of the major component, M_1A_1, ΔG is the Gibbs free-energy change of the reaction

$$M_1A_1 + M_2A_2 \rightleftharpoons M_1A_2 + M_2A_1,$$

X is the mole fraction of the minor component, M_2A_2, and L is the molar heat of fusion of the major component, M_1A_1.

Floquet's Theorem *See* **Bloch's Functions**.

Flürscheim's Theory of Benzene Substitution
Flürscheim suggested in 1902 a method of determining the orientation of benzene substituents which was developed subsequently by other workers. If the mono-substituent in a benzene ring is a positive group, then *m*-substitution takes place, whereas if it is a negative group *o–p* substitution occurs.

Fokker–Planck Equation
A description of the time-dependence of a **Markoff Process**. If $P(v, t/v_0)dv$ is the probability of a particle having a velocity between v and $v + dv$ at time t, when it is known to have a velocity v_0 at time $t = 0$, then

$$\frac{\partial P}{\partial t} + \frac{\partial}{\partial v}(M_1 P) - \tfrac{1}{2}\frac{\partial^2}{\partial v^2}(M_2 P) = 0$$

where M_1 and M_2 are given by

$$M_n = \frac{1}{\tau}\langle [\Delta v(\tau)]^n \rangle$$

Also $\langle[\Delta v(\tau)]^n\rangle = \langle[v(\tau)-v(0)]^n\rangle$ is the nth moment of the velocity increment in time τ, and τ is a time interval which is infinitesimal when considering the macroscopic system, so that any terms involving M_n where $n > 2$ can be neglected. Note that in deriving **Boltzmann's Transport Equation** a particle can change its velocity abruptly as a result of a collision, whereas here a macroscopic particle can only change its velocity by a small amount in time τ.

Foster's Reactance Theorem

The most general driving point reactance function representing a network, in which every mesh contains independent inductance and capacitance, has the form

$$Z = \mathrm{j}H\frac{(\omega^2-\omega_1^2)(\omega^2-\omega_3^2)\ldots(\omega^2-\omega_{2n-1}^2)}{\omega(\omega^2-\omega_2^2)(\omega^2-\omega_4^2)\ldots(\omega^2-\omega_{2n-2}^2)}$$

where H is a constant and the poles and zeros of the function are simple and alternate, i.e.

$$0 < \omega_1 < \omega_2 < \ldots < \omega_{2n-2} < \omega_{2n-1} < \infty$$

The theorem is useful in determining a network which will produce a certain specific current variation for a given source potential.

Foucault Current

A current induced in the interior of conductors by variations of magnetic flux.

Foucault's Pendulum

A freely suspended pendulum does not continue to swing in the same direction but appears to rotate in an opposite sense to the rotation of the earth. In fact, the plane of oscillation remains constant and it is the rotation of the earth about its axis which gives rise to the apparent rotation. ψ, the angle of rotation per hour, is given by

$$\psi = 15\sin\phi$$

where ϕ is the geographical latitude of the observer.

Fourier *See* **Ohm**, **Thermal** (**Appendix**)

Fourier–Bessel Integral

An expression corresponding to the **Fourier Integral**, but with exponential kernels replaced by **Bessel Functions**.

$$f(x) = \int_0^\infty J_m(kn)k\,dk \int_0^\infty f(\xi)J_m(k\xi)\xi\,d\xi$$

where J_m is a Bessel function of order m.

Fourier Integral
Fourier's Series becomes, in the limit for $-\infty < x < \infty$ and for continuous $m \to k$, the Fourier integral

$$f(x) = \frac{1}{2\pi}\int_{-\infty}^{\infty} e^{jkx}\,dk \int_{-\infty}^{\infty} f(\xi)e^{-jk\xi}\,d\xi$$

Fourier Number
A dimensionless number used in heat transmission and given by $\lambda t/C\rho l^2$ where λ is thermal conductivity, C specific heat, ρ density and l a characteristic length.

Fourier's Law (of Heat Conduction)
Heat flux in an isotropic medium is given by $-\lambda(dT/dx)$ where λ is the thermal conductivity and dT/dx is the temperature gradient.

Fourier's Series
Any function of x, say $f(x)$, can be expressed as the sum of a series of sines and cosines

$$\begin{aligned}&\tfrac{1}{2}b_0 + b_1 \cos x + b_2 \cos 2x + \ldots \\ &+ a_1 \sin x + a_2 \sin 2x + \ldots\end{aligned}$$

where

$$b_m = \frac{1}{\pi}\int_{-\pi}^{+\pi} f(x)\cos mx\,dx$$

and

$$a_m = \frac{1}{\pi}\int_{-\pi}^{+\pi} f(x)\sin mx\,dx$$

and this development holds for all values of x between $-\pi$ and $+\pi$ provided $f(x)$ is monotonic and is finite and continuous in the interval $-\pi < x < +\pi$, or, if discontinuous, has only finite discontinuities; a_m and b_m are called **Fourier coefficients**.

Fourier's series was developed in the process of finding a solution to an equation of the type

$$\frac{\partial^2 u}{\partial x^2} + \frac{\partial^2 u}{\partial y^2} = 0$$

which arose in a problem connected with the distribution of heat in a solid conducting medium.

Fourier Transform

If $f(x)$ is such that

$$\int_{-\infty}^{\infty} |f(x)|^2 \, dx$$

is finite, and if

$$F(k) = \frac{1}{\sqrt{(2\pi)}} \int_{-\infty}^{\infty} f(x) \, e^{jkx} \, dx,$$

then $F(k)$ is called the Fourier transform of $f(x)$. Also

$$f(x) = \frac{1}{\sqrt{(2\pi)}} \int_{-\infty}^{\infty} F(k) \, e^{-jkx} \, dk$$

Franck–Condon Principle

Electronic transitions in molecules occur in times which are very short in comparison with the period of vibration of the nuclei. Hence, in an energy level diagram, the most probable electronic transitions occur vertically.

Frankland's Method

Dialkyl zinc compounds readily react with alkyl halides to form hydrocarbons. The dialkyl zinc compounds are difficult to handle and this method is, in general, only used to form paraffins containing a quaternary carbon atom. Compare **Würtz Reaction**.

Franklin *See* **Appendix**

Franklin Equation
This relates the energy level of sound in a room as a function of time after the source has been cut off.

$$E = E_o \exp(-c\alpha St/4V)$$

where E and E_0 are the final and initial sound levels, c is the velocity of sound, S the surface area, V the volume of the room, t the time and α is the mean sound absorption coefficient.

Frank Partial Dislocation *See* **Shockley Partial Dislocation**

Frank–Read Source
A source of dislocation loops in a strained crystal. If a dislocation is pinned at two points (this can arise from two successive cross-slip processes), then the dislocation can bow out under stress through stages 2 to 4 until it produces the complete loop 5; whereupon the production of a new loop can begin.

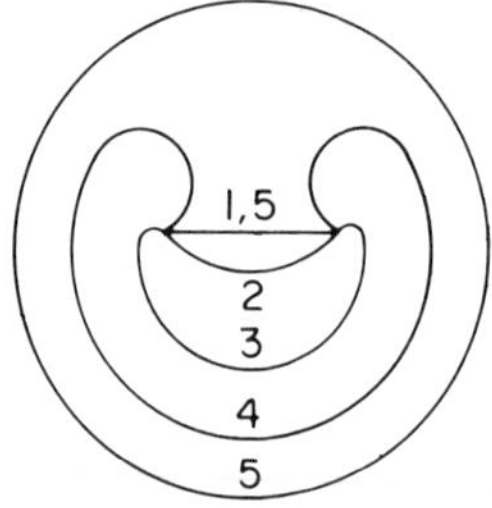

Frank–Read Source

Frank's Rule (for Dislocations)
If a perfect dislocation with **Burgers' Vector** $\mathbf{b}_1$ can dissociate into perfect dislocations with Burgers' vectors $\mathbf{b}_2$ and $\mathbf{b}_3$, then dissociation will occur only if

$$\mathbf{b}_1{}^2 < \mathbf{b}_2{}^2 + \mathbf{b}_3{}^2.$$

Franz–Keldysh Effect
A shift to longer wavelengths, in the absorption edge of semiconductors, produced by a strong electric field. Thus, optical transmission and reflectance can be modulated by the application of an electric field.

Fraunhofer *See* **Appendix**

Fraunhofer Diffraction
Diffraction phenomena may be divided into two classes:

I. When the light source and the screen upon which observations of the pattern are taken are at an infinite distance from the aperture, the diffraction is known as Fraunhofer diffraction.

II. When either the source or the screen, or both, are at a finite distance, then it is known as **Fresnel Diffraction**.

Fraunhofer Lines
Dark absorption lines found in the spectrum of the sun as observed from the surface of the earth and designated by letters A, B, C, . . . starting at the red end. It is customary to use light of the corresponding frequencies when specifying indices of refraction for optical materials.

Fredholm Equation
An integral equation of the same type as the **Volterra Equation** with constant limits of integration.

Freeth's Nephroid
A curve represented by a polar equation of the form

$$r = a\{1 + 2\sin(\theta/2)\}.$$

Nephroid means kidney-shaped, although this curve is somewhat more elaborate than kidney-shaped.

Frenet's Formulae *See* **Serret–Frenet Formulae**

Frenkel Defect

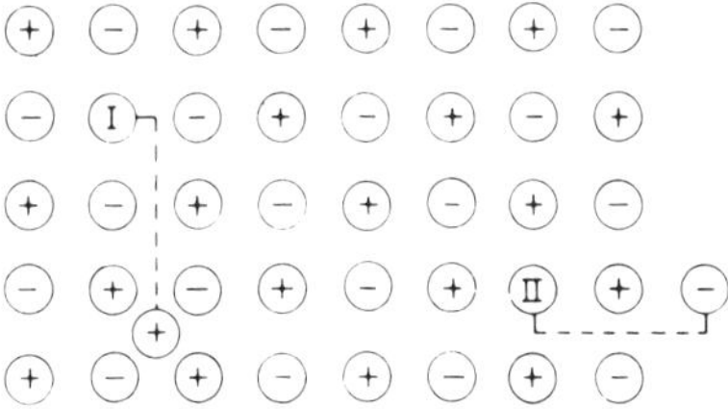

Frenkel Defect

There exists in a crystal in thermal equilibrium a number of vacant lattice points. An ion removed from a lattice position, and placed in an interstitial position in the lattice, leaves behind a Frenkel defect. If, however, the displaced ion is removed to the surface of the crystal, the defect is known as a **Schottky Defect**.

Frenkel Exciton
An exciton occurs in a crystalline solid and may be considered as a conduction-band electron and a valence-band hole bound together (although with finite separation) moving through the crystal. A Frenkel exciton is a tightly bound electron–hole pair with small separation of the electron and hole. A **Wannier** (or **Mott**) **Exciton** is a weakly bound pair with large separation.

Fresnel *See* **Appendix**

Fresnel–Arago Laws (of Polarized Interference)

I. Two rays polarized at right angles do not interfere.

II. Two rays polarized at right angles obtained from the same beam of polarized light will interfere in the same manner as ordinary light only when brought into the same plane.

III. Two rays, polarized at right angles and obtained from perpendicularly polarized components of unpolarized light, never interfere even after rotation of their planes of polarization.

Fresnel Diffraction *See* **Fraunhofer Diffraction**

Fresnel Integrals
The evaluation of the intensity of a diffraction pattern leads to an expression containing two integrals of the form:

$$\int \cos\frac{\pi v}{2}\,\mathrm{d}v \quad \text{and} \quad \int \sin\frac{\pi v}{2}\,\mathrm{d}v$$

where v depends upon the distances of the aperture from the source and from the screen and upon the displacement of the ray. These two integrals represent the components, along two rectangular axes, of the resultant amplitude; they may be evaluated between given limits by the use of tables calculated by Fresnel and other workers.

Fresnel Number
In an optical resonator with mirrors of radii a_1 and a_2 and separation d, the Fresnel number for light of wavelength λ is given by

$$N = a_1 a_2/\lambda d.$$

Loss from the resonator by diffraction will be small only if the Fresnel number is much larger than unity.

Fresnel Reflection Formulae

For incident light of intensity I_0 polarized with the magnetic vector in the plane of incidence, the intensity of light reflected from the surface of a transparent medium is

$$I = I_0 \frac{\sin^2(i-r)}{\sin^2(i+r)}$$

where i and r are the angles of incidence and refraction respectively. For light polarized with its magnetic vector perpendicular to the plane of incidence, the intensity of the reflected light is

$$I' = I_0 \frac{\tan^2(i-r)}{\tan^2(i+r)}.$$

For unpolarized light of incident intensity I_0 the reflected intensity is $\frac{1}{2}(I+I')$. For light polarized or unpolarized at normal incidence, passing from a medium of refractive index n to one of refractive index n', the intensity of the reflected light is

$$I_0\left(\frac{n'-n}{n'+n}\right)^2.$$

Fresnel Zone

In considering the effect of a wave front at any point P, it is convenient to separate the wave front into regions formed first by constructing a tangent sphere to the wave front with centre at P and radius, say, a.

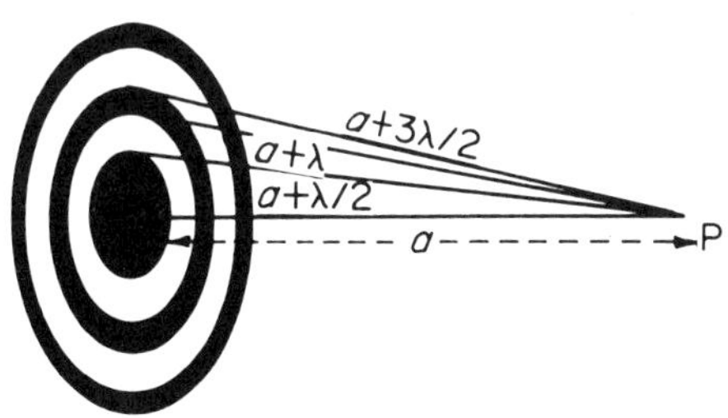

Fresnel Zone

Successive curves are then described on the wave front by spheres of radii $a+\frac{1}{2}\lambda$, $a+2\frac{1}{2}\lambda$, $a+3\frac{1}{2}\lambda$, etc., λ being the wavelength, and the zones enclosed between any two adjacent curves are called Fresnel zones. It can easily be shown that the areas of these zones are equal.

Freundlich Isotherm
The variation in adsorption with pressure at constant temperature may be represented by the equation

$$\frac{x}{m} = kp^{1/n}$$

where x is the mass of gas adsorbed by m grams of adsorbent at a pressure p, and k and n are constants for the system.

Freund Method (for Preparation of *Cyclo*paraffins)
When $\alpha\omega$ dihalogen derivatives of the paraffins are treated with sodium or zinc, *cyclo*paraffins are formed

$$CH_2\begin{cases} CH_2Br \\ CH_2Br \end{cases} + Zn \rightarrow CH_2\begin{cases} CH_2 \\ | \\ CH_2 \end{cases} + ZnBr_2$$

Compare **Würtz Reaction**.

Friedel–Crafts Reaction
This reaction is one of the most famous and useful reactions in organic chemistry. It consists of the condensation of an alkyl chloride, or an alkyl or aryl acid chloride, with an aromatic hydrocarbon in the presence of aluminium chloride. A simple example of the first reaction is the condensation of benzene with methyl chloride to yield toluene. A disadvantage of this reaction is, however, that polysubstitution can also occur; hence, in the presence of excess methyl chloride, benzene will be converted not only into toluene and xylene but in to the higher methyl benzenes, e.g. durene. Other anhydrous inorganic halides have been used as the catalyst but they are less efficient. The relative potencies of these catalysts are, for example:

$$AlCl_3 > FeCl_3 > SnCl_4 > BF_3 > ZnCl_2.$$

The ketone synthesis is free from the complicated side-reaction of the hydrocarbon synthesis and, in this case, the reaction between the acid chloride and the hydrocarbon occurs in the presence of a full equivalent of aluminium chloride. The reaction is usually exothermic to start with and the reactants are cooled, but the condensation is completed by refluxing the reactants together. The ketones produced by this reaction may be reduced to the hydrocarbon by means of the **Clemmensen Reduction**.

Friedel's Law
Every crystal diffracts x-rays as if a centre of symmetry were present. Thus the diffraction symmetry of a crystal is its point group symmetry plus a centre of symmetry, and corresponds to any one of eleven kinds of centrosymmetry (the **Laue Symmetry Groups**); i.e. the intensity of reflection from opposite sides of the same set of crystal planes is the same. The law can fail in polar crystals for radiation which is close in wavelength to an absorption edge.

Friedländer Synthesis
This synthesis is an important method of preparing quinoline and substituted quinolines: for example, when *o*-aminobenzaldehyde is condensed with acetaldehyde in the presence of aqueous sodium hydroxide, quinoline is formed.

Fries Rearrangement
Phenyl esters, when treated with anhydrous aluminium chloride, rearrange to give a mixture of *ortho*- and *para*-hydroxyketones. Generally low temperatures (60°C) favour the formation of the *p*-isomer, whereas high temperatures (above 160°C) favour the *o*-isomer.

Fries' Rule
The most stable arrangement of the bonds of a polynuclear compound is that form which has the maximum number of rings in the benzenoid form, i.e. three double-bonds in each ring.

Frobenius' Method
A method of solving homogeneous linear differential equations of any order, with variable coefficients, where the solution is assumed in the form of a series

$$y = x^{c}(a_0 + a_1 x + \dots + a_n x^n + \dots)$$

where c, a_0, a_1, . . . a_n, . . . are determined by substituting the series into the equation and setting the complete coefficient of any power of x to zero.

Froude's Number

A dimensionless number similar to the **Reynolds Number** for bodies falling through a fluid and given by

$$F = \frac{v^2}{la}$$

where v is the velocity, l a characteristic length and a the acceleration for each body.

G

Gabriel's Phthalimide Synthesis
Phthalimide is converted by means of alcoholic potassium hydroxide into the potassium salt. This salt, on heating with an alkyl halide, gives the *N*-alkyl phthalimide which may be hydrolysed by heating with 20 per cent hydrochloric acid under pressure or refluxing with potassium hydroxide solution to give phthalic acid and the *N*-alkyl amine.

Galilean Transformation
A transformation of **Cartesian Coordinates** x, y, z, t given by the equations

$$x' = x - vt$$
$$y' = y$$
$$z' = z$$
$$t' = t$$

where v is the velocity (in the x direction) with which one system of coordinates moves with respect to the other.

Galileo *See* **Appendix**

Galvano- Effects
Named after L. A. Galvani, these are effects occurring when an electric current is passed through a conductor (*see* **Thomson** and **Peltier Effects**). Similarly, galvanomagnetic effects occur when a magnetic field is also present (*see* **Hall** *and* **Ettingshausen Effects**).

Gamow Factor
An exponential factor e^{-G} which determines the yield of nuclear reactions. At low energies, G has the approximate form (in SI units) of

$$\frac{zZe^2}{2\varepsilon_0 \hbar v}$$

E

for a particle of charge ze approaching a nucleus of charge Ze at a velocity v.

Gamow–Teller Selection Rules (for Beta Decay) *See* **Fermi Selection Rules**

Gantmakher Effect
If a very thin plate of metal is placed in a magnetic field with the field parallel to the surface of the plate, and microwave radiation polarized perpendicular to the direction of the magnetic field is incident orthogonally on to the surface of the plate, then there will be resonant transmission through the plate for certain thicknesses or for certain values of the magnetic field. This resonant transmission corresponds to the thickness of the metal sheet equalling either the extremal orbital diameter of the electrons, as allowed from the shape of the **Fermi Surface** of the metal, or integral multiples of the diameter.

Gattermann Aldehyde Synthesis
When benzene is treated with a mixture of hydrocyanic acid and hydrochloric acid in the presence of aluminium chloride, a complex is formed which hydrolyses to benzaldehyde. The yield is poor. This synthesis is also applicable to phenols and phenolic ethers.

Gattermann Carbon Monoxide Synthesis of Aldehydes (Gattermann–Koch Reaction)
A mixture of carbon monoxide and hydrogen chloride, in the presence of anhydrous aluminium chloride and cuprous chloride, behaves like formyl chloride in reactions with aromatic compounds. The product of the reaction (which is only stable at low temperatures) is an aryl aldehyde. (*Compare* **Friedel–Crafts Reaction.**)

Gattermann Reaction
Chlorine or bromine may be introduced into the benzene or substituted benzene ring by diazotizing the appropriate amine in the presence of the acid and decomposing the diazonium salt with copper powder.

Gauss *See* **Appendix**

Gauss' Analogies *See* **Delambre's Analogies**

Gauss' Error Curve
In a set of observations where measurements are subject to random errors, the number of measurements associated with any given error may be expressed by a curve

$$y = \frac{h}{\sqrt{\pi}} e^{-h^2x^2}$$

where y is the number of observations with deviation x from the mean value and h is a constant giving the measure of precision of the set of observations. Such a set, if it follows closely the distribution given by the above relation, is termed a normal distribution.

Gauss' Hypergeometric Differential Equation

$$(x^2 - x)\frac{d^2y}{dx^2} + \left[(1 + \alpha + \beta)x - \gamma\right]\frac{dy}{dx} + \alpha\beta y = 0$$

with α, β and γ constants and γ not an integer. Series expansion about $x = 0$ gives a particular solution:

$$F(\alpha, \beta; \gamma; x) = 1 + \frac{\alpha\beta}{\gamma}x + \frac{1}{2!}\frac{\alpha(\alpha+1)\beta(\beta+1)}{\gamma(\gamma+1)}x^2 + \ldots$$

which converges uniformly for $|x| < 1$ and where $F(\alpha, \beta; \gamma; x)$ is the hypergeometric function.

Gaussian Brackets
A Gaussian bracket of n symbols is a polynomial which is linear in each of the symbols and the brackets satisfy the following relation:

$$[a_1 \ldots a_n] = [a_1 \ldots a_{n-1}] + [a_1 \ldots a_{n-2}]$$

with the initial conditions $[a_1] = a_1$ and $[\quad] = 1$. Use of Gaussian brackets has particular application to geometrical optics.

Gaussian Complex Integers
Complex numbers of the form $a + jb$, where a, b are rational integers and $j = \sqrt{-1}$.

Gaussian Curvature
At a general point on a surface there is a direction for which the radius of curvature of a normal section is a maximum, and a perpendicular

direction for which it is a minimum. The corresponding radii of curvature, ρ_1 and ρ_2 are called principal radii of curvature, and the total, or Gaussian, curvature is given by

$$\kappa = \frac{1}{\rho_1 \rho_2}$$

Gaussian Line Shape
When the dependence of amplitude on frequency for a wave group takes the form of **Gauss' Error Curve**, the line shape is said to be Gaussian.

Gaussian Positions
Any point on the axis produced of a bar magnet can be said to be in Gauss A position and any point on the magnetic equator can be said to be in Gauss B position.

Gaussian Units
A composite system of units in which all electric quantities are measured in absolute c.g.s. electrostatic units and all magnetic quantities are measured in c.g.s. electromagnetic units.

Gauss' Law
If a closed surface of any shape is constructed in an electric field *in vacuo*, the total electric flux crossing it in an outward direction is equal to $1/\varepsilon_0$ times the net positive charge through the surface, where ε_0 is the permittivity of free space. If the surface passes through a medium, then both free and induced charges enclosed by the surface must be taken into account; hence ε_0 must be replaced by ε, the permittivity of the medium.

Gauss' Multiplication Theorem

$$\Gamma(nz) = \sqrt{\left[\frac{n^{2nz-1}}{(2\pi)^{n-1}}\right]} \Gamma\left(z\right) \Gamma\left(z+\frac{1}{n}\right) \Gamma\left(z+\frac{2}{n}\right) \ldots \Gamma\left(z+\frac{n-1}{n}\right)$$

$$(n = 2, 3, \ldots)$$

See also **Euler's Definition of the Gamma Function.**

Gauss' Optics Formulae

I. For a single spherical surface: if s is the object distance, s' the image distance and n and n' the refractive indices of the two media separated by the spherical surface of radius of curvature r, then

$$\frac{n}{s}+\frac{n'}{s'}=\frac{n'-n}{r}$$

II. For a lens:

$$\frac{1}{s}+\frac{1}{s'}=\frac{1}{f}$$

where f is the focal length of the lens and there is the same medium on either side of the lens. (f is positive for a converging system, negative for a diverging one; if s is positive when measured to the left, s' is positive when measured to the right.)

Gauss' Π Function

$$\prod(k, z) = k^z \prod_{n=1}^{k} \frac{n}{z+n} \qquad \prod(z) = \lim_{k=\infty} \prod(k, z) = \Gamma(z+1)$$

Gauss' Reciprocal Theorem

Let charges $e_1, e_2, \ldots$ be placed at $P_1, P_2, \ldots$ and let V_1 be the potential at P_1 due to all charges except e_1. Let $e_1', e_2', \ldots$ be any other charges placed at the same points and let $V_1', V_2', \ldots$ be the corresponding potentials. Then

$$\Sigma eV' = \Sigma e'V$$

Gauss' Test (for Convergence or Divergence)

If for the series Σa_n the ratio of consecutive terms is given by

$$\frac{a_{n+1}}{a_n} = 1 - \frac{A}{n} + O\left(\frac{1}{n^{1+\lambda}}\right)$$

where $\lambda > 0$, then for $A > 1$, the series converges; if $A \leqslant 1$, the series diverges.

Gauss' Theorem

The surface integral of a vector over a closed surface is equal to the volume integral of the divergence of the vector throughout the enclosed volume.

$$\iiint \operatorname{div} \mathbf{A}\, \mathrm{d}V = \iint \mathbf{A}\, \mathrm{d}\mathbf{S}$$

Gay-Lussac Law of Combining Volumes
If gases interact and form a gaseous product, the volumes of the reacting gases and the volumes of the products are in simple proportion.

Gay-Lussac's Law (of Expansion)
At constant pressure, the coefficient of thermal expansion (volume coefficient) is the same for all gases and has a mean value

$$\alpha = 0{\cdot}0036609 = 1/273{\cdot}16.$$

Gegenbauer Polynomials
When n in the equation

$$(x^2-1)\frac{\mathrm{d}^2y}{\mathrm{d}x^2}+2(\beta+1)x\frac{\mathrm{d}y}{\mathrm{d}x}-n(n+2\beta+1)y=0$$

is zero or a positive integer, the solution becomes a polynomial (Gegenbauer polynomial); otherwise it is an infinite series. For $\beta = 0$, the differential equation becomes Legendre's equation satisfied by **Legendre's Coefficients**, and for $\beta = -0{\cdot}5$ the equation is satisfied by **Chebyshev Polynomials**. The Gegenbauer polynomials can be generated as coefficients of the power series

$$\frac{2^\beta}{(1-2tx+t^2)^{\beta+1/2}} = \frac{\sqrt{\pi}}{\Gamma(\beta+\frac{1}{2})}\sum_{n=0}^{\infty} t^n C_n^\beta(x)$$

where $\Gamma(\beta+\frac{1}{2})$ is the gamma function for $\beta+\frac{1}{2}$.

Geiger–Nuttal Rule
An empirical relationship between the radioactive constant λ and the range R of α-particles emitted by a radioactive element.

$$\log\lambda = b + c\log R$$

where b and c are constants, b differs for each of the three radioactive series and c is practically the same for each series.

Geitel Effect *See* **Elster and Geitel Effect**

Giauque's Temperature Scale
A temperature scale using only one fixed point, the triple point of

water, which is defined as 273·16 degrees kelvin. The scale was internationally accepted in 1954.

Gibbs *See* **Appendix**

Gibbs' Adsorption Equation

J. W. Gibbs deduced thermodynamically the relationship between surface tension and adsorption. If Γ is the surface concentration of solute per unit area of interface, a the activity of the solute and γ the surface tension, then

$$\Gamma = -\frac{1}{RT}\frac{\mathrm{d}\gamma}{\mathrm{d}(\ln a)} = -\frac{a}{RT}\frac{\mathrm{d}\gamma}{\mathrm{d}a}$$

Gibbs–Duhem Equation

A relation between the chemical potential and concentration of species in a mixture at constant temperature and pressure. If there are n_i moles of species i present having chemical potential μ_i, then

$$\sum_i n_i \,\mathrm{d}\mu_i = \sum_i x_i \,\mathrm{d}\mu_i = 0$$

where $x_i = n_i/\Sigma n_i$ is the mole fraction and $\mathrm{d}\mu_i$ is the change in chemical potential for a small change in the number of moles $\mathrm{d}n_i$.

Gibbs' Function (or Free Energy)

The Gibbs' function, or free energy, G, is given by

$$G = H - TS$$

where H is the enthalpy and S is the entropy.

Gibbs–Helmholtz Equation

This equation was deduced independently by J. W. Gibbs in 1875 and H. von Helmholtz in 1882. It may be expressed in a number of ways. If ΔG is the change in free energy during a reaction and ΔH the increase in heat content, then,

$$\Delta G - \Delta H = T\left[\frac{\partial \Delta G}{\partial T}\right]_P$$

Gibbs' Paradox

One might expect that, as two diffusing gases become more and more alike, the change in entropy due to diffusion should get smaller and smaller, approaching zero as the gases become identical. The fact that this is not the case is known as the Gibbs' Paradox. Bridgman explained it in the following fashion: in principle at least it is possible to distinguish the dissimilar gases by a series of experimental operations. In the limit, however, when the gases become identical there is a discontinuity in the operations in that no instrumental operation exists by which the gases may be distinguished. Hence a discontinuity in a function such as change in entropy is to be expected.

Gibbs' Phase Rule

$$F = C + 2 - P$$

where F is the number of degrees of freedom of a system (e.g. temperature, pressure or concentration) which must be fixed in order to define the system uniquely, C is the number of components, that is the smallest number of distinct substances which enable the constitution of each phase to be expressed, and P is the number of phases of the system, i.e. the homogeneous, mechanically separable, physically distinct portions of the heterogeneous system.

Gibbs' Phenomenon
In considering the transient response of an electrical network to a unit step function, the wider the band of frequencies passed by the network the better will be the response. Gibbs, however, stated that no matter how high is the cut-off frequency of the network, the maximum amplitude of the trailing overshoot of the leading edge never becomes less than 0·175. More generally, a **Fourier Series** expansion of a function round a discontinuity results in an overshoot, even if an infinite number of terms is taken.

Gibbs Ring *See* **Plateau Border**

Gibbs' Rule
J. W. Gibbs proposed a more rigid form of **Dalton's Law of Partial Pressures**. The pressure of a mixture of gases is only equal to the sum of the partial pressures if the chemical potential of the gas is unchanged by being incorporated in a mixture.

Gilbert *See* **Appendix**

Gilchrist–Thomas Process *See* **Bessemer Process**

Ginzburg–Landau Parameter
A macroscopic wave-function used in superconductivity theory to represent the superfluid component of the electron gas.

Giorgi System
A system of measurement in which the units are the metre, the kilogram, the second and the ampere.

Gladstone–Dale Law
The refractive index of a substance varies with change in volume according to the formula

$$k = \frac{n-1}{\rho}$$

where n is the refractive index, ρ the density and k is a constant.

Gödel's Theorems
I. If axiomatic set theory is consistent, there exist theorems which can neither be proved nor disproved.
II. There is no constructive procedure which will prove axiomatic set theory to be consistent.

Goldbach Conjecture
Any even number other than 2 can be represented as the sum of two prime numbers. It is unproved.

Goldschmidt's Law
After consideration of the crystal structures of a large numbet of inorganic compounds, V. Goldschmidt postulated in 1929 the fundamental law of crystal chemistry:

The structure of a crystal is determined by the ratio of the numbers, the ratio of sizes and the properties of polarization of its structural units.

Gomberg–Hey Reaction
Unsymmetrically substituted diaryls can be prepared by treating the appropriate aryl diazonium salt solution with sodium hydroxide or sodium acetate in the presence of a liquid aromatic compound.

Gomberg Reaction
When a solution of a diazonium salt is mixed with a liquid aromatic compound at low temperatures (5°C) and made alkaline with NaOH a small yield (10–40 per cent) of a product containing two united aromatic nuclei is obtained.

Graeffe's Method
This is a method of determining the roots of an equation of the form

$$\mathrm{f}(x) = a_0x^n - a_1x^{n-1} + \ldots a_n = 0.$$

A new equation is obtained of the form

$$\mathrm{f}(x)\mathrm{f}(-x) = 0$$

and the process repeated r times to obtain an equation whose roots are the 2^rth power of the roots of the original equation. Suppose that the roots of the original equation are $x_1, x_2, \ldots x_n$ where $x_1 > x_2 > x_3 \ldots > x_n$ and the equation of the nth degree in x^2 obtained by the squaring process is of the form

$$P_0z^n - P_1z^{n-1} + P_2z^{n-2} \ldots P_n = 0,$$

then

$$x_1{}^{2r} = \frac{P_1}{P_0}, \quad x_2{}^{2r} = \frac{P_2}{P_1}, \ldots x_n{}^{2r} = \frac{P_n}{P_{n-1}},$$

thus obtaining the roots for the original equation from the ratio of the coefficients in the derived equation. r is chosen for the accuracy desired. If the roots are complex or equal, then variations on the method are available.

Graetz Number
A dimensionless number used in problems of heat transfer and defined as

$$\frac{v\rho AC_p}{\lambda l}$$

where v is the velocity of flow, ρ the density, A the surface area, C_p the specific heat at constant pressure, λ the thermal conductivity and l a characteristic length.

Graham's Law of Diffusion
At constant temperature and pressure, the rate of diffusion of a gas is inversely proportional to the square root of its density.

Gram's Determinant

$$\begin{vmatrix} \mathbf{a}.\mathbf{a} & \mathbf{a}.\mathbf{b} & \mathbf{a}.\mathbf{c} \\ \mathbf{b}.\mathbf{a} & \mathbf{b}.\mathbf{b} & \mathbf{b}.\mathbf{c} \\ \mathbf{c}.\mathbf{a} & \mathbf{c}.\mathbf{b} & \mathbf{c}.\mathbf{c} \end{vmatrix} = (\mathbf{a}.\mathbf{b}\times\mathbf{c})^2$$

Grashof's Number
A dimensionless number, (Gr), for flow of a fluid past a body given by

$$(Gr) = \frac{\gamma T a l^3 \rho^2}{\eta^2}$$

where γ is the volume coefficient of thermal expansion for the fluid, a the acceleration, ρ the density, η the dynamic viscosity and l a characteristic linear dimension of the body.

Graves' Theorem

$$\mathrm{f}\left[\pi_1 + \frac{\mathrm{d}\phi}{\mathrm{d}\rho_1}, \pi_2 + \frac{\mathrm{d}\phi}{\mathrm{d}\rho_2}\right] \equiv \mathrm{e}^{\phi(\rho_1,\rho_2)}.\mathrm{f}(\pi_1,\pi_2).\mathrm{e}^{-\phi(\rho_1,\rho_2)}$$

where π and ρ are two distributive symbols of operation, combining according to the law

$$\rho.\pi \equiv \pi.\rho + a$$

Green's Dyadic
A vector operator corresponding to **Green's Function** when the appropriate differential equation is expressed in terms of vectors.

Green's Function
Green's function $G(x, t)$ is used to solve linear differential equations subject to boundary conditions. It is seldom used for numerical computation but is of fundamental importance for theoretical investigations. It is used, for example, for solving equations of the form

$$\mathscr{L}y(x) + \mathrm{f}(x) = 0$$

where $\mathscr{L}$ is a linear differential operator which does not depend explicitly on x and t, $G(x, t)$ will depend both on the form of th operator and on the boundary conditions which $y(x)$ must satisfy. A solution of the differential equation will be

$$y(x) = \int_a^b G(x, t)\, \mathrm{f}(t)\, \mathrm{d}t$$

provided $G(x, t)$ is correctly constructed. If $u(x)$ is a solution of the differential equation at $x = a$, and $v(x)$ is a solution at $x = b$, then

$$G(x, t) = \begin{cases} -\dfrac{1}{A} u(x)\, v(t) & x < t \\ -\dfrac{1}{A} u(t)\, v(x) & t > x \end{cases}$$

where A is a constant. The method can be generalized to applications involving higher dimensionality.

Green's Theorem

A relation expressing an integral taken over the surface of a number of bodies as an integral taken through the space between them.

If u, v, w are continuous functions of the Cartesian coordinates x, y, z, then

$$\Sigma \iint (lu + mv + nw)\,\mathrm{d}S = -\iiint \left(\frac{\partial u}{\partial x} + \frac{\partial v}{\partial y} + \frac{\partial w}{\partial z} \right) \mathrm{d}x\, \mathrm{d}y\, \mathrm{d}z$$

where Σ denotes that the surface integrals are summed over any number of closed surfaces and l, m, n are direction cosines of the normal drawn in every case from the element $\mathrm{d}S$ into the space between the surfaces. The volume integral is taken through the space between the surfaces.

In the special case, when

$$u = U\frac{\partial V}{\partial x}, \quad v = U\frac{\partial V}{\partial y}, \quad w = U\frac{\partial V}{\partial z}$$

Green's theorem becomes

$$\iiint (U\nabla^2 V - V\nabla^2 U)\,\mathrm{d}x\, \mathrm{d}y\, \mathrm{d}z = -\Sigma \iint \left(U\frac{\partial V}{\partial n} - V\frac{\partial U}{\partial n} \right) \mathrm{d}S$$

where $\partial/\partial n$ denotes differentiation along the normal to the surface S.

Gregory's Interpolation Formulae

These are obtained from **Newton's Interpolation Formula** when equal intervals are used. If $f(x)$ is given for $x_0, x_0+h, \ldots x_0+nh$, then

$$f(x) = f(x_0) + \frac{x-x_0}{h}\Delta f(x_0) + \frac{(x-x_0)(x-x_0-h)}{2!\,h^2}\Delta^2 f(x_0)$$

$$+ \ldots + \frac{(x-x_0)\ldots\{x-x_0-(n-1)h\}}{n!\,h^n}\Delta^n fx_0)$$

$$+ \frac{(x-x_0)\ldots(x-x_0-nh)}{(n+1)!}\left(\frac{d^{n+1}f(x)}{df(x)^{n+1}}\right)_{x=\zeta}$$

where $\Delta f(x_0) = f(x_0+h) - f(x_0)$

$$\Delta^2 f(x_0) = \Delta f(x_0) \text{ etc.,}$$

and ζ has a value intermediate between the greatest and least of x_0, x_0+nh and x. The formula is called **Gregory's forwards formula.** In addition **Gregory's backwards formula** applies for $f(x)$ given for x_0, $x_0-h, \ldots x_0-nh$, when

$$f(x) = f(x_0) + \frac{x-x_0}{h}\nabla f(x_0) + \frac{(x-x_0)(x-x_0+h)}{2!h^2}\nabla^2 f(x)\ldots$$

where $\nabla f(x_0) = f(x_0) - f(x_0-h)$ etc.
This form is often used for extrapolation.

Gregory's Series

$$\tan^{-1} x = x - \frac{x^3}{3} + \frac{x^5}{5} - \ldots + (-1)^{n-1}\frac{x^{2n-1}}{2n-1} + \ldots$$

that value of $\tan^{-1} x$ being understood which starts from zero with x.

Greninger Net

A net used for interpreting back-reflection Laue photographs. The net can be used for identifying groups of spots which arise as a result of reflections from groups of planes in the same zone, and for measuring angles between spots.

Grignard Reaction

The alkyl magnesium halides, Grignard reagents, are widely used in synthetic organic chemistry. The number of reactions in which a

Grignard reagent can take part is very great but, for brevity, most of the reactions may be classified under two heads:

(*a*) Reaction of the reagent with a group containing multiple bonds, for example a carbonyl, nitroso, isocyanide, cyanide or sulphone group. No reaction occurs with ethylenic or acetylenic bonds. The alkyl group adds on to the group with lower electron affinity whilst the MgX radical adds onto the other group, e.g.

$$R_1R_2CO + R_3MgX \rightarrow R_1R_2R_3COMgX.$$

This compound may then be hydrolysed or further treated.

(*b*) A Grignard reagent will react with compounds containing an active hydrogen group or a reactive halogen atom:

$$R.MgX + R'OH \rightarrow RH + (R'O)MgX.$$

The complex in both cases can be hydrolysed or used for further reactions.

Grotthus and Draper Law
Only radiations absorbed by a system are effective in producing chemical change.

Grotthus' Chain Theory
To explain the conductivity of electrolytes, von Grotthus, in 1805, made the assumption that the electrodes behaved as the poles of a magnet, the cathode attracting hydrogen and the anode attracting oxygen. He made no attempt to explain the nature of the forces involved, but imagined the molecules of the electrolyte stretched out in chains between anode and cathode, decomposition taking place on the molecules nearest to these electrodes.

Grove's Process
Alkyl chlorides may be produced by passing hydrochloric acid into the alcohol in the presence of anhydrous zinc chloride.

Grüneisen Constant or **Number**
The constant first appeared in the Grüneisen equation of state for solid bodies

$$pV + G(V) = \gamma E$$

where p is the pressure, $G(V)$ is the potential energy, V is the molar volume and $E = \int_0^T C_V \mathrm{d}T$ is the energy of atomic vibrations. (C_V is specific heat at constant volume.) γ is essentially a measure of the anharmonicity of the vibrations and is independent of temperature for most elements, lying between 1·5 and 2·5. For certain materials such as LiF and diamond, which have high Debye temperatures, γ drops with temperature.

Grüneisen Relation

The coefficient of thermal linear expansion α is related to the bulk modulus κ by the equation

$$\alpha = \frac{\gamma C_V}{3\kappa V}$$

where C_V is specific heat at constant volume and γ is the Grüneisen constant.

Gudden–Pohl Effect

The flash of light or luminescence which occurs in certain materials, such as zinc sulphide phosphors, when an electric field is applied or removed while the material is exhibiting phosphorescence (or afterglow). The effect is often called electro-photoluminescence.

Gudermannian

If $\cosh x = \sec \theta$ then θ is called the Gudermannian of x.

Guerbet Reaction

Higher branched-chain primary alcohols result when normal primary alcohols are heated with sodium ethoxide in the presence of a nickel catalyst.

Guillemin Effect

A magnetostrictive effect in which, if a bar of ferromagnetic material is elastically or permanently bent, it tends to straighten upon magnetization.

Guinier–Preston Zones

In the solid state, a phase of a material may contain microconstituents which do not show definite boundaries within the phase. These regions are referred to as Guinier–Preston zones.

Guldberg and Waage's Law
This law, known more commonly as the Law of Mass Action, states that, for a homogeneous system, the rate of a chemical reaction is proportional to the active masses of the reacting substances. The molecular concentration of a substance in solution or in the gas phase may be taken as a measure of the active mass.

Guldin's *or* **Guldinus' Theorems** *See* **Pappus' Theorems**

Gunn Effect
The occurrence of high-frequency (microwave region) current variations in certain semiconductors, notably gallium arsenide, on application of a high d.c. electric field. For a semiconductor to exhibit the effect, it must possess a negative change of electron mobility with electric field once the electric field has reached a threshold value. In gallium arsenide this decrease in mobility arises from the transfer of electrons from a high-mobility central valley in the conduction band to satellite valleys which are at a higher energy but where the mobility is lower. Under these conditions the charge distribution in the semiconductor sample becomes unstable, and there is established a high-electric-field domain which passes down the sample to the anode where it collapses. On formation of the domain the current in the sample falls, whereas on collapse of the domain the current increases again, allowing the nucleation of a new domain at the cathode. Hence a regular series of pulses is observed. This effect has technological application in the production of a simple semiconductor microwave generator.

Gurevich Effect
In an electrical conductor, over which a temperature gradient exists, phonons will themselves carry a thermal current (the lattice heat flow) and if phonon–electron collisions are important – as they are in a pure metal at low temperatures – the phonons may tend to drag the conduction electrons with them from hot to cold. This effect will produce an additional component of thermoelectric power.

Gyulai–Hartly Effect
An observed increase in electrical conductivity in ionic crystals when compressively deformed. The effect is thought to arise from electrical charges associated with dislocations in the crystals.

Haber Process

The reaction between nitrogen and hydrogen to form ammonia is exothermic and accompanied by a diminution in volume

$$N_2 + 3H_2 \rightleftharpoons 2NH_3 + \text{heat}$$

By **Le Chatelier's Principle** therefore, the reaction will be assisted by low temperatures and high pressures. There is, however, an optimum temperature on the practical scale, for although low temperatures will ultimately yield greater amounts of ammonia, the speed of reaction will be greatly reduced.

Hadamard's Inequality

If a determinant of order n has value D and complex elements a_{ij}, then

$$|D|^2 \leqslant \prod_{i=1}^{n} \sum_{j=1}^{n} |a_{ij}|^2$$

Hagens–Rubens Relation

This gives the reflectivity of metals in the far infrared region. Various forms of the relation are used but, for normal incidence, the reflection coefficient can be expressed in SI units in the form

$$\mathscr{R} = 1 - 2\left(\frac{2\omega\varepsilon_0}{\sigma_0}\right)^{1/2}$$

where ω is the angular frequency of the radiation, σ_0 is the electrical conductivity for static electric fields and ε_0 is the permittivity of free space. The expression is a good approximation at wavelengths greater than 30 μm.

Haidinger Brush

A yellowish brush, with a small blue cloud on either side, which can be seen by the eye by studying light reflected from a glass plate at the angle of polarization, or by studying the blue sky. The yellow brush always lies perpendicular to the direction of the electric vector of the light. The Haidinger brush is caused by dichroism of the yellow spot of the retina.

Haidinger Interference Fringes

Fringes produced by the interference of light reflected from, or transmitted by, two plane parallel surfaces of a thick transparent plate. The fringes are at infinity and not in the plane of the plate as in the case of very thin films.

Hall Effect

The Hall effect is a phenomenon which may be observed when conductors and semiconductors are subjected to an electric and a magnetic field orthogonal to each other. A voltage, the **Hall voltage** V_H, appears across the sample in a direction at right angles to both these fields. If the current density arising from the applied electric field is j_x, the magnetic flux density is B_z and the thickness is Δy then a **Hall coefficient** R can be defined:

$$V_H = Rj_x B_z \Delta y.$$

It can be shown for the case of electrons in a metal that the Hall coefficient is related to the carrier density n by the relationship

$$R = -\frac{1}{ne}.$$

For electrons in a pure semiconductor, because of a different velocity distribution, this relationship becomes

$$R = -\frac{3\pi}{8}\frac{1}{ne}.$$

For holes the sign is reversed, and hence the measurement of the Hall voltage is useful practically to distinguish between n- and p-type materials. If both carriers are present as in an intrinsic semiconductor, the relationship is more complicated

$$R = -\frac{3\pi}{8e}\frac{n\mu_n^2 - p\mu_p^2}{(n\mu_n + p\mu_p)^2}$$

where μ_n and μ_p are the Hall mobilities of electrons and holes respectively and n and p are their carrier densities.

The angle between the direction of current flow and the total electric field (i.e. the vector sum of the applied field and the Hall field) is called the **Hall angle**.

Hallwach's Effect
A negatively charged body in a vacuum is discharged by exposing it to ultraviolet light.

Hamiltonian

$$H = T + V$$

where T = kinetic and V = potential energy, is a Hamiltonian, if the total energy H can be expressed in terms of conjugate variables, i.e. momenta and coordinates only (and time if necessary).

If $T = p^2/2m$, where p is the momentum and m is the mass, the **Hamiltonian operator**, as used in quantum mechanics, becomes

$$H_{\text{op}} = -\frac{\hbar^2}{2m}\nabla^2 + V$$

where V is function of position.

Hamilton Jacobi Equation
This is the differential equation

$$H\left(\frac{\mathrm{d}S}{\mathrm{d}q}, q\right) = E$$

where H is the Hamiltonian of the system, q the generalized coordinates and S is a contact transformation function such that

$$\mathrm{d}S/\mathrm{d}q = p \qquad \text{(momentum)}.$$

Hamilton's Canonical Equations
In a generalized physical system with n degrees of freedom expressed in generalized coordinates $q_1, q_2, \ldots q_n$, and generalized momenta $p_1, p_2, \ldots p_n$, and with $H = H(q, \dot{q}, t)$, the Hamiltonian of the system, then

$$\dot{p}_i = -\frac{\partial H}{\partial q_i} \quad \text{and} \quad \dot{q}_i = \frac{\partial H}{\partial p_i}$$

($i = 1, 2, \ldots n$). *See also* **Lagrange's Equations of Motion**.

Hamilton's Principle

In a dynamical system composed of discrete material particles, where the kinetic energy T of the system is known as a function of the coordinates and their derivatives, and the potential energy V of the system is known as a function of coordinates and time, then the motion of the system is such that the integral

$$\int_{t_1}^{t_2} (T-V)(q_1, q_2, \ldots q_n; \dot{q}_1, \dot{q}_2, \ldots \dot{q}_n; t)\,\mathrm{d}t$$

is as small as possible. The q terms represent the generalized coordinates necessary to specify the configuration of the system.

Hammett Equation

An equation relating reaction rate constants to structural parameters. It is essentially a linear Gibbs free-energy relationship

$$\Delta G = \rho \Delta G_s$$

where ΔG represents the effect of a certain substituent on the free energy of activation, ΔG_s is the effect of that same substituent in an arbitrarily chosen standard reaction and ρ is a constant characteristic of the reaction being correlated.

Hammick–Illingworth Rule

If, in a benzene derivative C_6H_5XY, Y is in a higher group in the periodic table or, if in the same group is of lower atomic weight than X, then the second group to enter the nucleus goes into the *m*-position. In all other cases, including that in which XY is a single group, the group is *p*-directing. When the groups joining X and Y are the same $(-C=C-)$, the group XY is *o*–*p*-orientating. With mixed groupings the rule may be rigidly applied. Thus for $CHCl_2$, CH will be *o*–*p*-orientating and CCl will be *m*-directing and, as is found in practice, mixtures of the three substituents are formed.

Hankel Function

The Hankel functions are linear combinations of the ordinary solutions of Bessel's equation. Thus

$$\mathrm{H}_n^{(1)}(x) = \mathrm{J}_n(x) + \mathrm{j}\mathrm{N}_n(x)$$

$$\mathrm{H}_n^{(2)}(x) = \mathrm{J}_n(x) - \mathrm{j}\mathrm{N}_n(x)$$

where J_n and N_n are **Bessel Functions**.

Hankel's Integral

$$\frac{1}{\Gamma(x)} = \frac{\mathrm{j}}{2\pi} \oint_{C} (-t)^{-x} \mathrm{e}^{-t} \mathrm{d}t$$

The path of integration C starts at $+\infty$ on the real axis, circles the origin anticlockwise and returns to the starting point.

Hansen's Integral Formula
A Bessel function of order n can be represented by the integral equation

$$\mathrm{J}_n = \frac{1}{2\pi} \int_{-\pi}^{\pi} \mathrm{e}^{\mathrm{j}x\cos t} \mathrm{e}^{\mathrm{j}n(t-\pi/2)} \mathrm{d}t$$

$$= \frac{(-1)^n}{\pi} \int_{0}^{\pi} \mathrm{e}^{\mathrm{j}x\cos t} \cos nt \, \mathrm{d}t \quad (n = 0, 1, 2, \ldots)$$

Hantzsch Synthesis
A pyrrole derivative is formed when chloroacetone, a β-ketoester and a primary amine condense together.

Hardy–Schulz Rule
The flocculating power of ions increases rapidly with their charge.

Harker–Kasper Inequalities
Inequality relationships for the intensities of reflections of x-rays from different planes in a crystal. The inequalities arise from applying the **Cauchy–Schwartz Inequality** to expressions for structure factors as calculated for the different reflections. By giving upper limits to the magnitudes of structure factors, the inequalities are useful for obtaining the phases of a few strong reflections. The inequality relationships are different for different symmetry elements and need to be calculated for each space group.

Harkins' Rule
Atoms of even atomic numbers are more abundant in the universe than atoms of odd atomic numbers.

Harnack's Theorem
Let L be a simple contour and let S^+, S^- be the finite and infinite parts

into which the plane is divided by L (which does not itself belong to S^+ or S^-). Let $\mathrm{f}(t)$ be a real and continuous function on L. Then, if

$$\frac{1}{2\pi \mathrm{j}} \int \frac{\mathrm{f}(t)\,\mathrm{d}t}{t-z} = 0 \text{ for all } z \text{ in } S^+, \quad \mathrm{f}(t) = 0 \text{ everywhere on } L:$$

also if

$$\frac{1}{2\pi \mathrm{j}} \int \frac{\mathrm{f}(t)\,\mathrm{d}t}{t-z} = 0 \text{ for all } z \text{ in } S^-, \quad \mathrm{f}(t) = \text{constant on } L.$$

Hartmann Dispersion Formula
An expression giving the variation of refractive index n with wavelength

$$n = n_0 + \frac{A}{(\lambda - \lambda_0)^\alpha}$$

where n_0, A, λ_0 and α are constants for a given material.

Hartmann Test
To determine the aberration in a lens, a diaphragm containing a number of small apertures is placed before the lens and the course of the rays is plotted by photographing the pencils of light in planes on either side of the focus.

Hartree *See* **Appendix**

Hartree–Fock Technique
A method for calculating the effect of interactions between many bodies. It takes into account the exchange interaction between the bodies, although calculation of this interaction is laborious and usually requires approximations. The method has extensive application, particularly in solid-state and nuclear physics.

Hartree System of Units
A system of units in which the values of $\hbar$, the quantum unit of angular momentum, e, the charge on the electron in electrostatic units, and m, the mass of the electron, are set to unity. In this system the unit of length is $\hbar^2/me^2$, the radius of the first Bohr orbit for an electron moving non-relativistically around a proton of infinite mass, and the

unit of time is $\hbar^3/me^4$, the time taken for an electron to move in a Bohr orbit through 1 radian.

Hawking's Theorems

I. A stationary black hole must be either static (i.e. non-rotating) or axisymmetric.

II. In interactions involving black holes, the surface area of the event horizon can never decrease.

Haworth Synthesis

This preparation converts benzene into naphthalene and can be represented schematically as

CH_2CO, CH_2CO, O; $AlCl_3$; CO, CH_2, CH_2, COOH; Zn/Hg, HCl; CH_2, CH_2, CH_2, COOH; H_2SO_4; CH_2, CH_2, CH_2, C=O; Zn/Hg, HCl; CH_2, CH_2, CH_2, CH_2; Se

Heaviside Equations

Differential equations applicable to a transmission line:

$$-\frac{\partial V}{\partial x} = Ri + L\frac{\partial i}{\partial t}$$

$$-\frac{\partial i}{\partial x} = GV + C\frac{\partial V}{\partial t}$$

where R, L, G and C are the resistance, inductance, conductance and capacitance per unit length of line, i and V are current and voltage along the line and x and t are distance and time.

Heaviside Layer

Heaviside suggested that an ionized layer in the air, concentric with the earth's surface, might serve as a reflecting surface which would confine radiation between it and the earth. The existence of such a layer, or rather system of layers, has been conclusively proved. The two main levels have been given the names 'Heaviside' or 'Kennelly–Heaviside layer' for the lower layer and the '**Appleton Layer**' for the upper layer. More recent practice has been to call them the E and F layers respectively.

Heaviside–Lorentz System of Units
C.g.s. (centimetre, gram, second) units in which the force between two magnetic poles m_1 and m_2 a distance d apart in a medium of permeability μ is $m_1 m_2/\mu d^2$ and in which the force between two charges q_1 and q_2 a distance r apart in a medium of permittivity ε is $q_1 q_2/\varepsilon r^2$.

Heaviside's Expansion Theorem
Let

$$u = \frac{\mathrm{F}(p)}{\mathrm{f}(p)} u_0$$

where $\mathrm{F}(p)$ and $\mathrm{f}(p)$ are known functions of $p = \partial/\partial t$.
Heaviside's expansion theorem states that

$$u = u_0 \left\{ \frac{\mathrm{F}(0)}{\mathrm{f}(0)} + \sum \frac{\mathrm{F}(\alpha)}{\alpha \mathrm{f}'(\alpha)} \mathrm{e}^{\alpha t} \right\}$$

where α is any root (except 0) of $\mathrm{f}(p) = 0$, $\mathrm{f}'(p)$ is the first derivative of $\mathrm{f}(p)$ with respect to p, and the summation extends over all the roots of $\mathrm{f}(p) = 0$.
The solution reduces to $u = 0$ at $t = 0$.

Heaviside's Unit Function
A voltage which is zero for $t < 0$ and unity for $t \geqslant 0$.

Hefnerkerze (Hefner) *See* **Appendix**

Heisenberg Force
An exchange force between nucleons in which charge (equivalent to spin plus position) is exchanged. (*See also* **Majorana Force**.)

Heisenberg Uncertainty Principle
This principle, first put forward by W. Heisenberg in 1927, is one of

the fundamental bases of modern physics. The simultaneous precise determination of velocity or any related property (e.g. momentum or energy) of a material particle and its position is impossible. The smaller the particle the greater the degree of uncertainty. (*See* **Poisson's Bracket**.)

Heitler–London Covalence Theory
Lewis proposed, on empirical grounds, that a covalent bond corresponds to an electron duplet. Heitler and London, generalizing from a wave-mechanical treatment of the hydrogen molecule, postulated that such an electron duplet is formed by electrons having opposite spins, as it had been found by them that two hydrogen atoms can only combine when the spins of the two electrons are anti-parallel.

Hell–Volhard–Zelinsky Reaction
The usual method of preparing α-chloro or α-bromo acids is to treat the acid with chlorine or bromine in the presence of red phosphorus.

$$\mathrm{R.CH_2.COOH} \xrightarrow{\mathrm{P+Br_2}} \mathrm{RCH_2CO.Br} \rightarrow \mathrm{R.CH.Br.CO.Br}$$

Helmholtz *See* **Appendix**

Helmholtz Double Layer
Electrokinetic phenomena (e.g. endosmosis) were explained by G. Quincke in 1859 as due to a charged layer at the liquid–solid interface. This concept of the existence of differently charged layers or electrical double layers was extended by H. von Helmholtz in 1879, who suggested that a double layer is invariably formed at the boundary between two phases. If the electrical double layer is regarded as an electrical condenser of parallel plates a molecular distance apart, the subject may be treated mathematically.

Helmholtz Function (or Free Energy)

$$F = U - TS$$

where F is the Helmholtz function or free energy, U is the internal energy, S is the entropy and T the absolute temperature.

Helmholtz's Equation (for a Reversible Electrolyte Cell)
If E is the e.m.f. of a reversible cell and ΔU is the change in internal

energy when nF coulombs pass through the cell, then

$$E = -\frac{\Delta U}{nF} + T\left(\frac{\partial E}{\partial T}\right)_p.$$

Helmholtz's Equation (Partial Differential Equation)
A partial differential equation

$$\nabla^2\psi + k^2\psi = 0$$

obtained from the wave equation when the time-dependence is harmonic.

Helmholtz's Relation (Optics)
If α_1 and α_2 are the inclinations of light rays to an axis passing through a spherical surface, where the refractive indices on the respective sides of the surface are n_1 and n_2, then the object and image heights y_1 and y_2 are related by

$$y_1 n_1 \tan\alpha_1 = y_2 n_2 \tan\alpha_2.$$

The equation is only valid for small angles and may be approximated by

$$y_1 n_1 \alpha_1 = y_2 n_2 \alpha_2.$$

Usually in this form it is referred to as **Lagrange's law**.

Helmholtz's Theorem *See* **Thévenin's Theorem**

Helmholtz's Theorem (for Fluids)
In the absence of body forces (when entropy is a constant), individual vortices in a non-viscous fluid always consist of the same fluid particles.

Henderson Equation (for Continuous Mixture Boundaries)
The combination of an electrode of unknown potential with a reference electrode often involves contact between two solutions of different electrolytes. A liquid junction potential is introduced therefore into the observed e.m.f. of the cell. P. Henderson, from the assumption that the boundary between two different electrolytes consists of a series of mixtures of the electrolytes in all proportions,

was able to deduce an equation for the potential.

$$E_B = \frac{RT}{F} \frac{c(u_+ + u_-) - c'(u_+' + u_-')}{c(u_+z_+ - u_-z_-) - c'(u_+'z_+' + u_-'z_-')} \ln \frac{c(u_+z_+ + u_-z_-}{c'(u_+'z_+' + u_-'z_-')}$$

where c and c' are the concentrations of the two electrolytes, u and u' are the speeds of the ions and z and z' are the valencies of the ions. The subscripts refer to anions and cations.

Henderson Equation for pH

In 1908 L. J. Henderson deduced that the pH of an acid during neutralization may be represented by

$$\mathrm{pH} = \mathrm{pK_a} + \log \frac{\text{concentration of salt}}{\text{concentration of acid}}$$

where $\mathrm{pK_a}$ is the logarithm to base 10 of the reciprocal of the dissociation constant of the acid. This equation is applicable for solutions between pH 4 and pH 10, provided they are not too dilute.

Henrici's Notation *See* **Bow's Notation**

Henry; Henry, Thermal *see* **Appendix**

Henry's Law

The mass of gas dissolved by a given volume of solvent at a constant temperature is proportional to the pressure of gas with which the solvent is in equilibrium.

Hermann–Mauguin Symbols

A notation used in describing the symmetry classes of crystals. *See* **Schoenflies Crystallographic Notation.**

Hermite Polynomials

Solutions of the Hermite differential equation

$$\frac{d^2y}{dx^2} - 2x\frac{dy}{dx} + 2ny = 0$$

when n is an integer.

The polynomials are given by

$$H_n(x) = (2x)^n - \frac{n(n-1)}{1!}(2x)^{n-2} + \frac{n(n-1)(n-2)(n-3)}{2!}(2x)^{n-4} + \ldots$$

or by the coefficients in the expansion

$$e^{2xs - s^2} = \sum_{n=0}^{\infty} \frac{H_n(x)s^n}{n!} \qquad |s| < \infty$$

The Hermite polynomials give the probability density function (the wave-function) of a simple harmonic oscillator in quantum theory.

Hermitian Matrix
If $\mathbf{A}$ is a square matrix, then the Hermitian adjoint $\mathbf{A}^+$ is given by $(\mathbf{A}^+)_{mn} = (\mathbf{A})^*_{nm}$ where rows and columns in $\mathbf{A}$ have been interchanged and then the complex conjugate of each element taken. A Hermitian matrix is a matrix equal to its own Hermitian adjoint i.e.: $\mathbf{A}^+ = \mathbf{A}$. A skew-Hermitian or anti-Hermitian is given by: $\mathbf{A}^+ = -\mathbf{A}$.

Hermitian Operator
A quantum-mechanical operator, say P, such that

$$\int_\tau u^*.Pv\,d\tau = \int_\tau v.P^*u^*\,d\tau$$

where * denotes conjugate quantity and the integration is over the whole space.

Heron's (or Hero's) Formula
The area of a triangle of sides a, b and c is given by

$$A = \sqrt{[s(s-a)(s-b)(s-c)]}$$

where $s = (a+b+c)/2$.

Hertz *See* **Appendix**

Hertz Effect
The enhancement of a spark discharge by passing ultraviolet light to promote ionization.

Hertzian Waves
Electromagnetic waves of frequencies between zero and approximately 10^4 MHz.

Hertzsprung–Russell Diagram
A plot of visual magnitude against colour index for stars grouped in particular galactic clusters. These axes are equivalent to luminosity versus temperature.

Hertz Vector
This is a single vector for an electromagnetic wave, from which both the electric and magnetic fields can be obtained. The vector is given by

$$\Pi = \int \mathbf{A}\, dt$$

where **A** is the vector potential.

Herzig–Meyer Method
Methyl-imino groups may be quantitatively estimated by heating the compound with hydrogen iodide at 150°C.

Hessian
If

$$y_k = \frac{\partial F}{\partial x_k} \quad (k = 1, 2, \ldots n)$$

where F is a function of $x_1, x_2, x_3, \ldots x_n$, then the symmetrical determinant

$$H = \frac{\partial^2 F}{\partial x_i, \partial x_j}$$

is the Hessian.

Hess' Law
The heat change in a chemical reaction is the same whether the reaction takes place in one or in several stages.

Heurlinger Equations
Equations relating adiabatic and isothermal thermoelectric and thermomagnetic effects. (In an adiabatic effect there is no transverse

flow of heat, whereas in an isothermal effect there is no transverse temperature gradient.) Thus if subscripts a and i represent the adiabatic and isothermal coefficients respectively:

$$R_a = R_i + \Sigma_i P$$
$$Q_a = Q_i - \Sigma_i S$$
$$\rho_a = \rho_i - Q_i P B^2$$
$$\lambda_a = \lambda_i (1 + S^2 B^2)$$
$$\Sigma_a = \Sigma_i + Q_i S B^2$$

where R is the Hall coefficient, P the Ettingshausen coefficient, Σ the thermoelectric power, ρ the electrical resistivity, λ the thermal conductivity, Q the Nernst coefficient, S the Righi–Leduc coefficient and B the magnetic flux density. These equations are sometimes referred to as the **Callen Relations**. (*See also the* **Hall, Ettingshausen, Nernst, Righi–Leduc** and **Maggi–Righi–Leduc Effects**.)

Heusler Alloys
Ternary metal alloys which contain manganese, aluminium and copper but no ferromagnetic element yet which exhibit ferromagnetism. Other similar ternary ferromagnetic alloys also exist.

Hilbert Space
A complete linear vector space with a complex-type scalar product.

Hilbert Transform
If

$$\mathrm{H}(z) = \frac{1}{\pi} \mathscr{P} \int_{-\infty}^{\infty} \frac{\mathrm{h}(y)}{y - z} \mathrm{d}y$$

then

$$\mathrm{h}(y) = -\frac{1}{\pi} \mathscr{P} \int_{-\infty}^{\infty} \frac{\mathrm{H}(z)}{z - y} \mathrm{d}z$$

are Hilbert transforms of each other. $\mathscr{P}$ indicates that the principal parts of the integrals should be taken.

Hill's Determinant
A determinant occurring in the solution of Mathieu's equation (*see* **Mathieu's Function**). The zeros of this determinant establish the stability of the solution.

Hiltner–Hall Effect
The polarization of light from distant stars. It is thought to occur in interstellar regions and to be due to the selective absorption of light vibrating in one plane by ferromagnetic grains which are aligned in a magnetic field in these regions.

Hinsberg Amine Separation
A mixture of primary, secondary and tertiary amines may be separated into its components by reaction with *p*-toluene sulphonyl chloride $CH_3.C_6H_4.SO_2Cl$. This reagent produces a 1:1 adduct with the primary and secondary amines and does not react with the tertiary amine. The reaction product with the primary amine can be separated from that of the secondary amine, as it is soluble in potassium hydroxide. (*See also* **Hofmann Amine Separation.**)

Hittorf Dark Space *See* **Crookes Dark Space**

Hoesch Reaction
If a nitrile is used instead of hydrogen cyanide in the **Gattermann Aldehyde Synthesis**, a ketone is formed. This modification is known as the Hoesch reaction.

Hofmann Amine Separation
The mixture of the three amines is heated with ethyl oxalate. Tertiary amines do not react; primary and secondary amines react according to the following schemes:

I. $$\begin{matrix} COOC_2H_5 \\ | \\ COOC_2H_5 \end{matrix} + 2RNH_2 \rightarrow \begin{matrix} COONHR \\ | \\ COONHR \end{matrix} + 2C_2H_5OH$$

Crystalline

II. $$\begin{matrix} COOC_2H_5 \\ | \\ COOC_2H_5 \end{matrix} + R_2NH \rightarrow \begin{matrix} COONR_2 \\ | \\ COOC_2H_5 \end{matrix} + C_2H_5OH$$

Liquid

The reaction mixture is then distilled. The first fraction contains the tertiary amine, the second fraction the diamide, whilst the residue in the distillation flask contains the monoamide. The amines themselves are regenerated from the amides by hydrolysing with alkali and distilling off the amine.

Hofmann Degradation

This reaction results in the conversion of an amide into a primary amine with one carbon atom less, by the action of bromine or fluorine and an alkali. The reaction can be carried out using the amides of all monocarboxylic acids and with yields of up to 90 per cent for acids having up to seven carbon atoms.

Hofmann Exhaustive Methylation Reaction

The thermal decomposition of quaternary ammonium compounds is a most important reaction. Tetramethylammonium hydroxide is the only compound of this type which decomposes to give an alcohol.

$$(CH_3)_4NOH \rightarrow (CH_3)_3N + CH_3OH$$

All other quaternary ammonium halides give an olefin and water.

$$(C_2H_5)_4NOH \rightarrow (C_2H_5)_3N + H_2O + C_2H_4$$

If the compound contains different alkyl groups, one of which is methyl, this group remains attached to the nitrogen. Further, if an ethyl group is present, ethylene is formed preferentially to any other olefin (**Hofmann rule**). Water is preferentially eliminated between the hydroxyl group and a β-hydrogen atom in one of the alkyl groups. This reaction is the basis of the Hofmann exhaustive methylation reaction, which is used to determine the position of a nitrogen atom in a heterocyclic ring, rather than to prepare unsaturated compounds

Hofmann Isonitrile Synthesis

Isonitriles are formed by the reaction of primary amines with chloroform in the presence of an alkali. The obnoxious smell of the poisonous isocyanide or carbylamine is used as a test for primary amines, as the other types of amine are not reactive under these conditions.

Hofmann Mustard Oil Reaction

Alkyl isothiocyanates may be prepared by heating together a primary amine, carbon disulphide and mercuric chloride.

Hofmann Rearrangement
This reaction may be used to prepare the homologues of aniline. For example, when phenyltrimethylammonium chloride is heated under pressure, the following reaction occurs.

$C_6H_5N(CH_3)_3Cl \xrightarrow[\text{Pressure}]{300^\circ C} N(CH_3)_2HCl\text{-}C_6H_4\text{-}CH_3 \longrightarrow NHCH_3HCl\text{-}C_6H_3(CH_3)\text{-}CH_3 \longrightarrow NH_2HCl\text{-}C_6H_2(CH_3)_2\text{-}CH_3$

Hofmann Rule *See* **Hofmann Exhaustive Methylation Reaction**

Hofmeister Series
A list of anions and cations arranged in the order of their increasing effectiveness in causing the swelling of gelatine.

Hölder Condition
Let $f(t) = f_1(t) + jf_2(t)$ be some complex function of the point t, of L. If, for every pair of points t_1 and t_2 of L

$$|f(t_2) - f(t_1)| \leqslant A|t_2 - t_1|^{\mu}$$

where A and μ are positive constants, $f(t)$ is said to satisfy the Hölder condition. A and μ are called the **Hölder constant** and **index**.

Hölder's Inequality
If f_1 and f_2 are two functions

$$\left|\int_a^b f_1 f_2 \, dx\right| \leqslant \left[\int_a^b |f_1|^p dx\right]^{1/p} \left[\int_a^b |f_2|^{p/(p-1)} \, dx\right]^{(p-1)/p}$$

$$(1 \leqslant p < \infty)$$

The inequality implies a similar inequality for sums of convergent series:

$$\left|\sum_j a_j b_j\right| \leqslant \left[\sum_j |a_j|^p\right]^{1/p} \left[\sum_j |b_j|^{p/(p-1)}\right]^{(p-1)/p}$$

The inequality reduces to the **Cauchy–Schwarz Inequality** if $p = 1$. *See also* **Minkowski's Inequality**.

F

Hooke's Law
Within the elastic limit of any body, the ratio of stress to strain is constant. As originally formulated by Hooke, the law stated that the ratio of force to extension is constant.

Hôpital *See* **L'Hôpital**

Hopkins–Cole (Glyoxylic Acid) Reaction
A violet ring obtained when concentrated sulphuric acid is added to a mixture containing a protein and glyoxylic acid.

Hopkinson's Coefficient
The ratio of the mean flux per turn of an induction winding to the mean flux per turn of another winding linked with it.

Horner's Method (for Determining the Real Roots of an Equation)
Let

$$\mathrm{f}(x) = a_0 x^n + a_1 x^{n-1} + \ldots a_n x + a_n = 0$$

be the equation and let p_1 be a first approximation to the root x_1 sought, such that $p_1 + 1 > x_1 > p_1$. Diminish the roots of the equation by p_1. In the transformed equation

$$a_0(x - p_1)^n + a_1(x - p_1)^{n-1} + \ldots + a_{n-1}(x - p_1) + a_n = 0$$

put

$$\frac{p_2}{10} = \frac{a_n}{a_{n-1}}$$

and diminish the roots by $p_2/10$ yielding

$$b_0\left(x - p_1 - \frac{p_2}{10}\right)^n + b_1\left(x - p_1 - \frac{p_2}{10}\right)^{n-1} + \ldots + b_n = 0.$$

Then take

$$\frac{p_3}{100} = \frac{b_n}{b_{n-1}}$$

and continue the operation. The required root will be

$$x_1 = p_1 + \frac{p_2}{10} + \frac{p_3}{100} + \ldots$$

If, however, b_n and b_{n-1} are of the same sign, p_2 was taken too large and must be diminished.

Houben–Hoesch Synthesis
This preparation is an extension of the **Gattermann Aldehyde Synthesis** but is not applicable to phenol itself. Cyanides condense with polyhydric phenols (particularly *m*-compounds) in the presence of zinc chloride and hydrogen chloride to give phenolic ketones.

Hubbard Model
This is a highly simplified model which attempts to account for magnetic ordering in metals. In the model the large number of bound and continuum electron levels associated with each ion is reduced to a localized orbital level. The level of each ion is specified as empty, or containing one electron of one or other spin, or two electrons of opposite spins, and the Hubbard Hamiltonian derived on this basis. This is found to contain two types of terms. The first type gives rise to local magnetic moments; the second type gives rise to a band spectrum. When both types of terms are present, exact analysis is not possible, but it can predict in certain cases a transition from a non-magnetic metal to an antiferromagnetic insulator. The latter is sometimes referred to as a **Mott–Hubbard Insulator**.

Hubble Effect
Distant galaxies show, in their spectra, redshifts which increase proportionally with the distance of the galaxy. This effect is consistent with the hypothesis of an expanding universe.

Hubble's Law
The ratio of velocity to distance of galaxies in the universe is a constant. The constant is called **Hubble's constant**. Its reciprocal is the time calculated to have elapsed since the collapsed state of the universe existed and is called **Hubble time**. This time is thought to be approximately $1{\cdot}9 \times 10^{10}$ years. The Hubble constant may increase at extreme distances because of a decrease of recessional velocities under the influence of gravity.

Hugoniot Function *See* **Rankine–Hugoniot Relations**

Hume-Rothery Rules

Hume-Rothery pointed out that particular alloy phases often occur at the same ratio of valence electrons to atoms.

Phase	*Electrons/Atoms*
α(face-centred cubic)	1·36–1·42
β(body-centred cubic)	1·48–1·50
γ(complex)	1·58–1·67

Hund's Rule

A rule enabling the relative order to be obtained of the different energy states for a configuration of equivalent electrons: i.e. when the electrons have the same quantum numbers n (principal quantum number) and l (orbital angular momentum) and thus form a sub-shell in an atom. Hund's rule states that the configuration which has the largest multiplicity or S value (where S is the quantum number for resultant spin of the electrons) is the lowest in energy, and if there are several of these terms, then the lowest of these has the highest value of L (quantum number for resultant orbital momentum). This means that there will be the maximum number of unpaired electrons consistent with filling the lowest energy levels first.

Hunsdieker Reaction

If a silver carboxylate is allowed to react with an equivalent quantity of bromine in carbon tetrachloride at boiling point, the alkyl halide is produced. The reaction provides a method of reducing the length of a carbon chain by one carbon atom.

Hüttig Equation

An equation used to express gas adsorption by non-porous solids. If V is the volume of gas adsorbed at pressure p and temperature T, and V_m is the volume of gas required to cover the surface completely with a monomolecular layer, then

$$\frac{V}{V_m} = \frac{cx(1+x)}{1+cx}.$$

x is the ratio of the equilibrium pressure p to the saturated gas pressure p_s of the adsorbent at temperature T. c is given by

$$c = A \exp\{(q - q_L)/RT\}$$

where q is the heat of adsorption onto a first-layer molecule, q_L is the heat of liquefaction of the absorbent and A is a constant. The equation gives adsorption values which are too small at higher relative pressures.

Huygens' Principle (Huygens–Fresnel Principle)
Every point of a wave can be considered as the centre of a new elementary wave. The resultant wave produced by the interference of all these elementary waves is identical with the original wave.

Hylleraas Coordinate
A coordinate system which leads to a separable solution of the **Schrödinger Equation** for the two-particle system, e.g. helium atom. The coordinates are $v = r_{12}$, $s = r_1 + r_2$, $t = r_1 - r_2$, where r_1 and r_2 are the distances of two electrons from the nucleus and r_{12} is the inter-electron distance.

I

Ilkovik Equation

In polarography, the limiting diffusion current in microamperes of a substance reduced or oxidized at a dropping mercury electrode is given by:

$$i_{\mathrm{d}} = 607nD^{1/2}Cm^{2/3}t^{1/6}\left(1+\frac{AD^{1/2}t^{1/6}}{m^{1/3}}\right)$$

where n is the number of electrons transferred for each of the molecules reacted, D is the diffusion coefficient in $\mathrm{cm}^2\,\mathrm{s}^{-1}$, C is the concentration in millimoles, m is the rate of flow of the mercury in $\mathrm{mg\,s}^{-1}$ and t is the time in seconds. A is a constant whose value is typically in the range 17–39. The numerical factor 607 contains the **Faraday constant**, a series of geometric factors and the conversion factors for the units chosen. The original Ilkovik equation did not contain the expression in the brackets; it was added later to take into account the deviation from sphericity of the expanding droplet of mercury.

Ising Model

Used, for instance, in ferromagnetism and phase-transformation theory, it is a model for a system of interacting particles. A scalar variable σ associated with each lattice point of a crystal has possible values $+1$ or -1 representing the two directions of electron spin in a ferromagnet or the two types of atom in a binary alloy. The interaction energy is given by

$$E_{ij}\begin{cases} = -J\sigma_i\sigma_j & \text{where } i, j \text{ are nearest-neighbour particles} \\ = 0 & \text{where } i, j \text{ are not nearest neighbours.} \end{cases}$$

J is a coupling constant which can be positive as in a ferromagnetic system, or negative as in an antiferromagnetic system. The model enables a value for the heat capacity to be calculated; the two-dimensional form and the three-dimensional form (which has not been solved exactly) lead to a transition point.

Ivanov Reagent
A reagent analogous to a Grignard reagent (*see* **Grignard Reaction**):

$$\begin{array}{l} C_6H_5.\underset{|}{C}H.COO\,Mg\,Br. \\ \quad\quad\;\; Mg\,Br \end{array}$$

It reacts with carbon dioxide to give phenyl malonic acid, with iodine to give substituted succinic acids, and with aldehydes and ketones to produce α-phenyl β-hydroxy acids.

J

Jacobian
If

$$y_k = f_k(x_1, x_2, x_3, \ldots x_n)$$

the determinant:

$$J = \begin{vmatrix} \frac{\partial y_1}{\partial x_1} & \frac{\partial y_1}{\partial x_2} & \cdots & \frac{\partial y_1}{\partial x_n} \\ \frac{\partial y_2}{\partial x_1} & \frac{\partial y_2}{\partial x_2} & \cdots & \frac{\partial y_2}{\partial x_n} \\ \cdots & \cdots & \cdots & \cdots \\ \frac{\partial y_n}{\partial x_1} & \frac{\partial y_n}{\partial x_2} & \cdots & \frac{\partial y_n}{\partial x_n} \end{vmatrix}$$

is the Jacobian.

Jacobi Polynomials
The Jacobi polynomials are defined as

$$J_n(a, c, x) = F(a+n, \ -n, \ c, \ x)$$

where F is the hypergeometric function and n is an integer. They are solutions of the differential equation

$$(x^2 - y^2)\frac{d^2y}{dx^2} + \left[(1+a)x - c\right]\frac{dy}{dx} + n(a+n)y = 0$$

See also **Gauss' Hypergeometric Differential Equation.**

Jacobi's Elliptic Functions *See* **Legendre's Elliptic Integrals**

Jacobi's Identity

$$[X, [Y, Z]] + [Y, [Z, X]] + [Z, [X, Y]] = 0$$

where [ ] is **Poisson's Bracket**.

Jacobi's Theta Functions
If v is a complex variable and $q = e^{j\pi t}$ is a complex parameter such that τ has a positive imaginary part, then the four theta functions are given by

$$\vartheta_1(v) = 2 \sum_{n=0}^{\infty} (-1)^n q^{(n+1/2)^2} \sin\{(2n+1)\pi v\}$$

$$\vartheta_2(v) = 2 \sum_{n=0}^{\infty} q^{(n+1/2)^2} \cos\{(2n+1)\pi v\}$$

$$\vartheta_3(v) = 1 + 2 \sum_{n=1}^{\infty} q^{n^2} \cos(2n\pi v)$$

$$\vartheta_4(v) = 1 + 2 \sum_{n=1}^{\infty} (-1)^n q^{n^2} \cos(2n\pi v)$$

Jacobsen Rearrangement
During sulphonation, polyalkylbenzenes or halogenated polyalkylbenzenes fairly readily undergo isomerization, due to the migration of the alkyl group or the halogen. The reaction may be intra- or intermolecular and both migrations have been shown to occur at the same time. The alkyl group migrates to the vicinal position. This reaction has not been found to occur in compounds containing:

$$-NH_2, \; -NO_2, \; -OCH_3 \text{ or } -COOH.$$

Jahn–Teller Effect
The electron states of a non-linear polyatomic molecule cannot show degeneracy in the ground state: i.e. there cannot be more than one possible quantum-mechanical state of the same energy. Thus, if degeneracy of the electron states due to symmetry of the molecule should exist, the atoms in the molecule will be displaced to destroy this symmetry.

Japp–Klingermann Reaction
Diazonium salts react with monoalkyl-substituted acetoacetic esters to produce the aryl hydrazone of the α-keto acid.

Jeans Length
The critical wavelength at which oscillations in an infinite homogeneous medium become gravitationally unstable and become a

stable bound system independent of the medium. The wavelength is given by

$$\lambda_J = c_s\,(\pi/G\rho)^{1/2}$$

where c_s is the isothermal value of the speed of sound in the medium, G is the gravitational constant and ρ is the density of the medium. The wavelength is of the order of 10^{18} m.

Joffé Effect
An observed decrease in brittleness of rock salt on dissolution of the surface. The effect is probably associated with the dislocations in the crystal.

Johnsen–Rahbek Effect
If a piece of semiconducting material is placed in contact with two electrodes, the frictional force between the electrodes and the semiconductor increases with increase of applied voltage to the electrodes. The effect is additional to any increase of friction that may occur due to electrostatic attraction of the electrodes. Similar variation of the shear resistance of a fluid with variation of voltage applied to two electrodes immersed in the fluid is called the **Winslow Effect**.

Johnson Noise
In electronics, noise arising from thermal agitation of electrons in conductors.

Jones Symbols
A notation for expressing symmetry operations within a point group. The symbols express the new positions of the x, y and z axes after the operation. For instance, a 180° rotation about the x axis is expressed by the symbol $x\bar{y}\bar{z}$.

Jones Zone
This is similar to, and in certain cases equal to, the **Brillouin Zone** for a solid; and the distinction is not always made. It is a zone for which an energy gap exists at all points on its surface. The surface of the zone generally consists of partial Brillouin zone boundaries from several zones.

Josephson Effect
If two superconductors are connected by a thin layer of insulating material with no electrostatic potential present, there is a quantum-mechanical penetration, through the layer, of electrons existing as **Cooper Pairs**. The arrangement is called a Josephson junction and the tunnelling is known as the **d.c. Josephson effect.**

If a steady potential difference V is applied, an oscillatory current passes through the layer of frequency

$$\omega = 2\pi\left(\frac{e'V}{h}\right)$$

where e' is the effective charge of the Cooper pairs. This is known as the **a.c. Josephson effect.**

Joshi Effect
A fall (negative effect) or rise (positive effect) in the low-frequency alternating current passing through a gas dielectric condenser when the gas is irradiated continuously with visible light.

Joule *See* **Appendix**

Joule Effect
A rod of para- or diamagnetic material suffers a mechanical change in length when subjected to a magnetic field. *See also* **Joule's Law**.

Joule's Law
In a linear conductor, that is one whose resistance R is a constant independent of total electric current I, the rate of development of heat is equal to RI^2.

Joule–Thomson Coefficient *See* **Joule–Thomson Effect**

Joule–Thomson Effect (Joule–Kelvin Effect)
When a gas is allowed to expand adiabatically through a porous plug, the temperature of the gas changes. The rate of change of temperature with pressure on adiabatic expansion, known as the **Joule–Thomson coefficient**, may be taken to be constant if the pressure difference across the plug is small. The differential Joule–Thomson effect, $\Delta T/\Delta p$, where ΔT is the change in temperature for a small change in

pressure Δp at temperature T, can be shown from thermodynamical considerations to be given by the expression

$$\Delta T = \frac{T[\partial V/\partial T]_p - V}{C_p} \Delta p$$

where V is the volume of the gas and C_p is the specific heat at constant pressure. The sign of the Joule–Thomson coefficient will depend upon the relative values of the terms in the numerator of the above expression. The value of the numerator is dependent upon the temperature, and there is a fixed temperature at which the sign of the Joule–Thomson coefficient changes. This temperature is known as the Joule–Thomson inversion temperature. Most gases are below their inversion temperatures at room temperature and hence may be liquefied by expansion. The so-called permanent gases (e.g. hydrogen, $T_i = 190$ K), have an inversion temperature below room temperature and must be cooled before they can be liquefied by this means.

Jurin Rule

The height of rise of a liquid in a capillary tube of radius a is given by:

$$h = \frac{2\gamma \cos \alpha}{\rho g a}$$

where γ is the surface tension of the liquid, α is the angle of contact with the capillary and ρ is the density of the liquid.

K

Kaiser Effect
Very low-energy acoustic pulses, which are observed when a metal is initially deformed, and which are not observed on reapplication of any stress up to the original limit. The effect arises from work-hardening removing the plastic region of the stress–strain curve. Acoustic emission reappears upon reversal of the stress or upon application of a stress exceeding the original limit. *See also the* **Bauschinger Effect**.

Kane Band
A form of energy band which commonly occurs in semiconductors which have a narrow gap between the valence and conduction bands. For such bands, if the energy of the electrons (or holes) is expressed as a function of wave-vector, the bands are very 'skinny'. The effective mass is a function of the electron or hole energy in the band and is very small for small electron (or hole) energies.

Kapitza Effect
The temperature discontinuity which occurs at the surface of a solid when heat flows from the solid into the liquid. The effect is of particular importance when the liquid is helium 4 or helium 3.

Kapp Line *See* **Appendix**

Kayser *See* **Appendix**

Kekulé Benzene Formula
In 1872 Kekulé postulated that benzene was a tautomeric mixture of

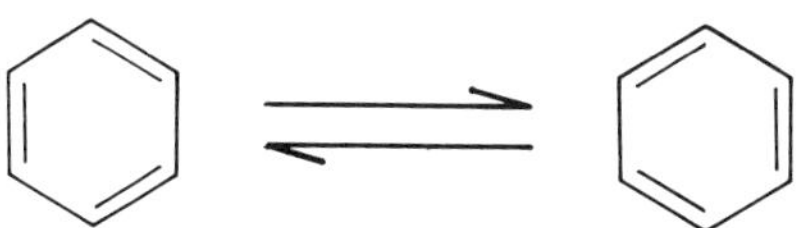

in order to explain the equivalence of the six hydrogen atoms Quantum-mechanical calculation has shown that the ground state of such a system leads to a coplanar structure with interacting p orbital giving rise to a π bond shared between all six carbon stoms; hence the modern symbol

See also **Ladenburg Benzene Formula**

Kelvin Effect *See* **Thomson Effect**

Kelvin's Equation
This relates the Peltier coefficient to the thermoelectric power for a conductor (or semiconductor).

$$\pi(B) = T\Sigma(B)$$

where $\pi(B)$ is the Peltier coefficient which is a function of magnetic flux density B if the conductor is placed in a magnetic field, and $\Sigma(B)$ is the thermoelectric power. (*See also* **Peltier** *and* **Nernst Effects**.)

Kelvin Skin Effect
A non-uniform distribution of variable currents in solid conductors, resulting in an increase in current density near the surface.

Kelvin's Statement (of the Second Law of Thermodynamics)
No process is possible whose sole result is the abstraction of heat from a reservoir and the performance of an equivalent amount of work: i.e. it is impossible to construct a perfect heat engine. (*See* also **Clausius' Statement**.)

Kelvin's Theorem
For an ideal homogeneous fluid in a region of flow where the entropy is constant, the circulation around a closed path moving with the fluid remains constant: i.e. vorticity cannot be created or destroyed under these conditions.

Kelvin Temperature Scale
Practical temperature scales usually depend upon the expansion of a particular substance and, since the coefficients of expansion are not constant, a scale defined in this fashion will be subject to some uncertainty. W. Thomson (Lord Kelvin) suggested a temperature scale based on the efficiency of a reversible machine. The zero of the Kelvin scale is the temperature of the sink of such a machine working at unit efficiency, i.e. converting heat completely into work. This is only possible at absolute zero on the gas scale of temperature and hence, provided that the degree is defined in the same fashion, these two scales are identical. By definition since 1954, 0 K is $-273{\cdot}16°$ C.

Kennard Packet
The minimum uncertainty state function $\chi(\mathbf{r})$ which can be used to describe a particle. It is given by

$$\chi(\mathbf{r}) = [2\pi(\Delta\mathbf{r})^2]^{-1/4} \exp\left[\frac{-(\mathrm{r} - \langle \mathrm{r} \rangle)^2}{4(\Delta\mathbf{r})^2}\right].$$

Thus the particle is centred at position $\langle \mathbf{r} \rangle$ with uncertainty $\Delta\mathbf{r}$. The corresponding uncertainty in momentum space is obtained by changing all $\mathbf{r}$ to $\mathbf{p}$ where $\mathbf{p}$ is the momentum. Any arbitrary decrease in $\Delta\mathbf{r}$ would increase $\Delta\mathbf{p}$, causing the wave packet to spread out again with time.

Kennedy's Theorem
Any three bodies having plane motion relative to each other have only three *centros* which lie on a straight line. A *centro* is either a point common to two bodies having the same velocity, or a point on a body about which another body turns or tends to turn.

Kennelly–Heaviside Layer *See* **Heaviside Layer**

Kepler's Equation
An expression for motion in an ellipse;

$$\theta - \varepsilon \sin\theta = \frac{2\pi t}{T}$$

where θ is the eccentric angle, ε the eccentricity and T the period.

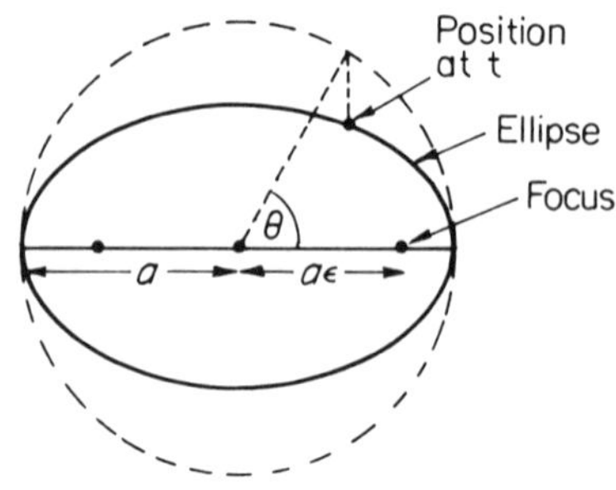

Kepler's Equation

Kepler–Poinsot Polyhedra
These are four regular polyhedra additional to the five **Platonic Polyhedra** and possessing re-entrant angles.

Kepler's Laws
I. Every planet moves in an ellipse of which the sun occupies one focus.
II. The radius vector drawn from the sun to the planet sweeps out equal areas in equal times.
III. The squares of the times taken to describe their orbits by two planets are proportional to the cubes of the major semi-axes of the orbits.

Kerr Electro-optic Effect
J. Kerr discovered in 1875 that certain materials, when placed in a stationary electric field, become optically anisotropic and doubly refracting. If the electric field is applied perpendicular to the direction of the light, then the effect is called the Kerr effect. If n_1 and n_2 are the refractive indices for light whose planes of polarization are respectively parallel and perpendicular to the electric field E, then

$$n_1 - n_2 = k\lambda E^2$$

where λ is the wavelength of the light and k is Kerr's constant. This effect has become of technological importance in optical communication systems, but see also the **Pockels Effect**.

Kerr Magneto-optic Effect
Light, plane polarized in or normal to the plane of incidence, becomes

elliptically polarized when reflected from the polished pole of an electromagnet.

Ketteler–Helmholtz Formula

In media which exhibit anomalous dispersion, the refractive index n for light of wavelength λ is given by

$$n^2 = 1 + \sum \frac{A_m \lambda^2}{\lambda^2 - \lambda_m{}^2}$$

if the medium contains sets of molecules with different periods of vibration. λ_m is the wavelength in free space of light of the same frequency as the natural frequency of the mth type of molecule, A_m is a constant referring to that molecule and the summation is taken to include all the natural periods of vibration of the molecules in the substance. Alternatively, it is sometimes referred to as the **Sellmeyer Formula**.

Khintchine's Theorem

If $x_1, x_2, \ldots x_n$ is a sequence of statistically independent random variables which all have the same probability distribution with mean value ζ, then as $n \to \infty$

$$x = \frac{1}{n}(x_1 + x_2 + \ldots x_n)$$

converges in probability to ζ.

Kikuchi Lines

Black or white lines which appear in the electron diffraction patterns of highly perfect crystals. When an electron beam is incident on the crystal, some electrons are diffusely scattered with loss of energy. These electrons then can be reflected back in the direction of the plate used to exhibit the diffraction pattern. A series of these processes gives rise to white (weakened) lines and black (enhanced) lines on the plate, these lines being parallel to and equidistant from the projection of each crystal plane.

Kiliani Reaction

The sugar series may be ascended by this reaction. Thus if an aldopentose is treated with hydrocyanic acid, a cyanhydrin is formed which, on hydrolysis with barium hydroxide and acidification with the correct quantity of dilute sulphuric acid, gives rise to a polyhydroxyacid. The solution is evaporated to dryness and the γ-lactone is formed which, on reduction with sodium amalgam in slightly acid solution, yields the aldohexose.

Kirchhoff's Approximation
A solution of the integral equation for scattering problems, involving the substitution of the incident wave for the unknown scattering density function.

Kirchhoff's Equation for Heat of Sublimation
An equation of the form

$$L_s = \int_0^T C_p^{\mathrm{f}} \mathrm{d}T - \int_0^T C_p^{\mathrm{i}} \mathrm{d}T + L_0$$

where C_p^{i} and C_p^{f} are the specific heats at constant pressure at the initial and final states and L_0 is the heat of sublimation at absolute zero.

Kirchhoff's Equations
The relationship between the variation of heat of reaction ΔH with temperature and the change of heat capacity ΔC_p at constant pressure p during a process is given by

$$\Delta C_p = \left[\frac{\partial(\Delta H)}{\partial T}\right]_p$$

Similarly if ΔU is the internal energy change, then the change of heat capacity ΔC_V at constant volume is given by

$$\Delta C_V = \left[\frac{\partial(\Delta U)}{\partial T}\right]_V$$

Kirchhoff's Law of Emission
The emissive power divided by the absorption coefficient, for any substance, depends only on the frequency and plane of polarization of the radiation and on the temperature, and is independent of the nature of the substance.

Kirchhoff's Laws of Current

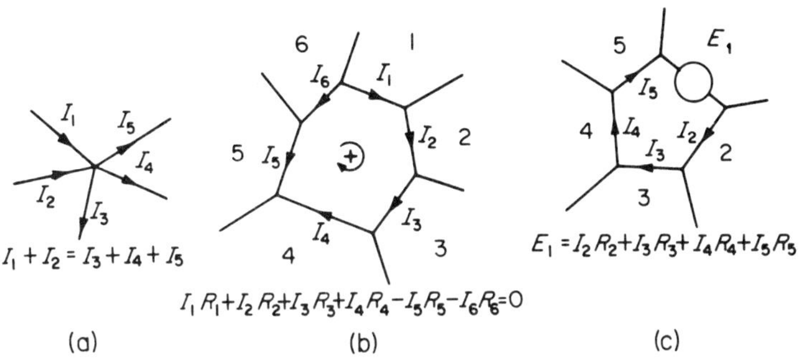

Kirchhoff's Laws of Current

In any circuit composed of a network of conductors:

(1) The algebraic sum of the currents at any junction of conductors (branch point) must be zero: $\Sigma I = 0$ (*see diagram* (a)).

(2) $\Sigma IR = \Sigma E$. The left-hand side sums the products of the currents in, and the resistances of, parts of the conductors forming a closed circuit in the network; the right-hand side is the sum of all the discontinuities of potential met in passing round the circuit. The proper signs must be given to the I and E terms throughout (*see diagrams* (b) *and* (c)).

Kirkendall Effect

If inert marker wires are embedded at the common face of a copper and a brass (CuZn) block, which is held at a high temperature, the wires marking the original interface move into the brass. This is because the zinc atoms diffuse past the interface faster than the copper atoms. On condensing, excess vacancies can cause the formation of voids in the brass. The effect, discovered by F. C. Kirkendall (1942), confirmed in the above experiment by Smigelskas and Kirkendall (1947) and since shown in many other systems, provides proof of the predominance of a vacancy mechanism for diffusion in pure metals and substitutional alloys.

Klein Bottle

This is a closed surface constructed so that it has one side and no boundary. The surface must intersect itself.

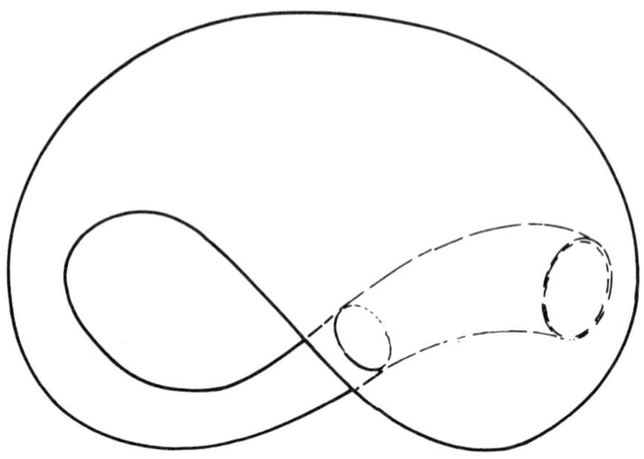

Klein Bottle

Klein–Gordon Equation
This is a differential wave equation of the form

$$\frac{1}{c^2}\frac{\partial^2\psi}{\partial t^2}=\frac{\partial^2\psi}{\partial x^2}-\mu^2\psi$$

and can be used to describe the motion of a flexible string in an elastic medium, electromagnetic waves in a dispersive medium, or the motion of a series of coupled pendulums. The constants c and μ can be obtained for the particular application, c being a velocity.

The equation holds for the de Broglie waves of relativistic free particles and, in particular, can be used to describe the 'scalar' meson, i.e. one without spin. However, in the presence of spin the equation becomes the **Proca Equation**. *See also* **De Broglie's Theory**.

Klein–Nishina Formula
The scattering cross-section per electron for Compton scattering into solid angle $d\Omega$ is given for unpolarized radiation by

$$d\sigma_C=\frac{r_0{}^2}{2}\left(\frac{E'_p}{E_p}\right)^2\left(\frac{E_p}{E'_p}+\frac{E'_p}{E_p}-\sin^2\theta\right)d\Omega$$

where r_0 is the classical radius of the electron.
For remaining notation *see* **Compton Effect**.

Knight Shift
W. D. Knight (1949) discovered that the nuclear magnetic resonance frequency in metals (or alloys) is higher than it is in chemical compounds containing nuclei of the same isotopes in the same magnetic field. The effect is due to the small extra field at the nucleus produced by electrons in the conduction band.

Knoevenagel Reaction
If an aldehyde is treated with an equivalent amount of diethylmalonate in the presence of pyridine and the product hydrolysed and heated, an α, β-unsaturated acid is formed. In practice it is usually sufficient to treat the aldehyde with malonic acid in the presence of pyridine.

Knoop's β-Oxidation Theory
Compounds containing an aromatic nucleus are not so readily oxidized by organisms as those containing straight chains.

Knorr Synthesis
Pyrrole derivatives may be prepared by the condensation of an α-amino ketone with a β-diketone or β-ketoester.

Knudsen Cosine Law
Kinetic theory shows that, for a gas at rest and at a constant temperature, the number of molecules $\mathrm{d}n$ striking or leaving an area $\mathrm{d}A$ of the wall subtending a solid angle $\mathrm{d}\phi$ at an angle θ to the normal is given by

$$\mathrm{d}n = \frac{\mathrm{d}A}{4\pi} n\bar{c} \cos\theta \, \mathrm{d}\phi$$

where n is the number of molecules per unit volume and $\bar{c}$ is the mean velocity of the molecules.

Knudsen Flow
That flow which occurs when the mean free path of the molecules is greater than the dimensions of the apparatus. The ratio of the latter to the former is often called **Knudsen's number**.

Knudsen Formula for Gas Flow
The mass of gas q passing any point in a tube per second at a certain

temperature and very low pressure is given by

$$q = \frac{(2\pi\rho)^{1/2}}{6} d^3 \frac{p_1 - p_2}{l}$$

where $(p_1 - p_2)/l$ is the pressure gradient along the length of the tube l, d is the diameter of the tube and ρ is the density of the gas at the given temperature and unit pressure.

Kohlrausch Law (of Independent Migration of Ions)
Each ion contributes a definite amount to the total conductance of the electrolyte, irrespective of the nature of the other ion. This rule is strictly true only at infinite dilution where there is no ionic attraction between the ions, i.e.

$$\Lambda_0 = \lambda_+ + \lambda_-$$

where λ_+ and λ_- are the ionic conductances at infinite dilution and Λ_0 is the equivalent conductivity.

Kohlrausch Square Root Law
In the calculation of equivalent conductance of electrolytes, the extrapolation of Λ_C, the equivalent conductance at concentration C, to Λ_0, the equivalent conductance at zero concentration, is made easier by plotting Λ_C as ordinate against $\sqrt{C}$ as abscissa, the relationship being linear. (*See* **Onsager Conductivity Equation**.)

Kohn Anomaly
When the lattice vibrational spectra of heavier metals such as lead are plotted, there is a kink in the frequency–wave-vector curve referred to as the Kohn anomaly. Only phonons (lattice vibrations) with a restricted range of wave-vectors can interact with electrons and the kink occurs where this interaction ceases.

Kohler's Rule
For normal magnetoresistance, the change of resistivity $\Delta\rho$ when a magnetic field H is applied to a conductor can be represented in the form

$$\frac{\Delta\rho}{\rho} = \mathrm{f}\left(\frac{H}{\rho_0}\right)$$

where ρ_0 is the resistivity in zero magnetic field. ρ_0 is inversely proportional to the relaxation time τ, so that the rule is equivalent to expressing magnetoresistance as a function of $H\tau$.

Kolbe Reaction

Most aliphatic acid salts, upon electrolysis with a smooth platinum electrode, give rise to a hydrocarbon and carbon dioxide at the anode.

Kolbe Synthesis

This is the original method of preparing salicylic acid. If sodium phenoxide is heated with carbon dioxide at 180–200° C, the following reaction occurs.

$$2\,C_6H_5ONa + CO_2 \rightarrow C_6H_4(ONa)COONa + C_6H_5OH$$

This reaction was modified by Schmitt (**Kolbe–Schmitt synthesis**). Sodium phenoxide is heated at 120–140°C under pressure with carbon dioxide and totally converted to the sodium salt of salicylic acid.

$$C_6H_5ONa + CO_2 \rightarrow C_6H_4(OH)COONa$$

Kondo Effect

Dilute metal alloys containing transition elements show a resistance minimum when resistance is expressed as a function of temperature, and this is referred to as the Kondo effect. Whereas the ideal resistance of the host metal varies as T^5 at low temperatures, the resistance due to the magnetic impurities increases with decreasing temperature. A full treatment of the effect requires advanced quantum mechanics.

Konowaloff's Rule

The vapour from a mixture of two liquids is relatively richer in the component whose addition to a liquid mixture results in an increase of total pressure.

Kopp's Law
The specific heat of a solid element is the same whether free or combined. In general this is not true.

Körner's Absolute Method (for Assigning the Position of Substituents in Benzene)
This method is based on the principle that the introduction of a third substituent into a *p*-isomer gives one trisubstituted product; into an *o*-isomer two such products; and into an *m*-isomer three trisubstituted isomers.

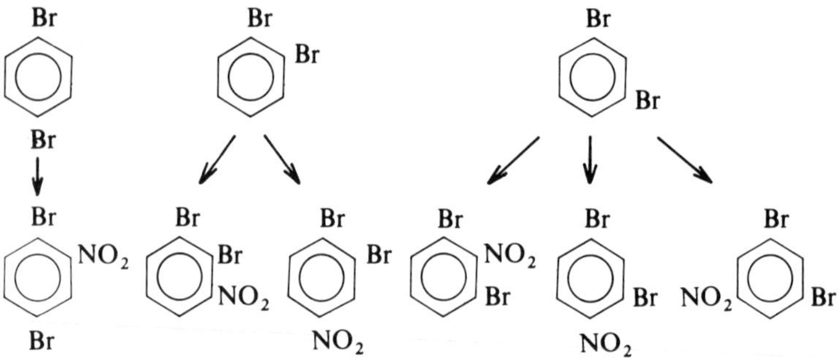

In practice this method is difficult, as the separation of the isomers may be almost impossible when present in minute amounts. The reverse reaction is better. Thus if each of the six diamino-benzoic acids is heated with soda lime, three phenylenediamines are formed. Three of the acids give the same diamine, which must threfore have the *m*-structure, whereas the diamine given by two of the acids must be the *o*-isomer and the other diamine must have the *p*-structure.

Kossel Lines
These are analogous to **Kikuchi Lines** in electron diffraction and occur when a point source of x-radiation is used. The x-rays meet the atomic planes at a continuous range of angles and those striking at the Bragg angle (*see* **Bragg's Law**) are reflected. There will be a cone of such rays and these give rise to a black arc (Kossel line) on the photographic film, either for back-reflection or transmission depending on the angle of the cone. Corresponding to the cone of diffracted rays there will be also a deficiency cone which can give rise to a white arc in the transmission pattern.

Kossel–Sommerfeld Law
The arc spectrum of an element having an odd atomic number shows even multiplicity, whereas the arc spectrum of an element having an even atomic number shows odd multiplicity.

Köster Effect
Plastic deformation of a material lowers the elastic modulus by introducing free dislocations.

Kramers–Krönig Relations
Relationships in dispersion theory which relate the frequency variation of the real and imaginary parts of response functions such as susceptibility. If $\chi_1(\omega_0)$ and $\chi_2(\omega_0)$ are the real and imaginary components of susceptibility in a medium at frequency ω_0, then one form of the relations is

$$\chi_1(\omega_0) = \mathscr{P}\int_0^\infty \frac{2\omega\chi_2(\omega)}{(\omega_0{}^2 - \omega^2)}\,\mathrm{d}\omega$$

and

$$\chi_2(\omega_0) = \mathscr{P}\int_0^\infty \frac{2\omega_0\chi_1(\omega)}{(\omega^2 - \omega_0{}^2)}\,\mathrm{d}\omega$$

where $\mathscr{P}$ means the principal part of the integral, that is integration for all frequencies ω excluding the vanishingly small region around ω_0. Proof of the expressions requires integration in the complex plane. The dielectric constant and the refractive index for a medium can be calculated from the susceptibility.

Kramer's Theorem
In electron paramagnetic resonance in solid materials, a twofold spin degeneracy is not removed by electric fields or spin-orbit coupling if the spin is half integral.

Kröger–Vink Diagram *See* **Brouwer Diagram**

Kronecker Delta
An operator defined as follows:

$$\delta_{nm} = 1\ (m = n)$$
$$= 0\ (m \neq n)$$

Kronecker Product

In group theory, the Kronecker (or direct) product of two three-dimensional representations of a group is obtained by multiplying the characters in the two representations. The character of a representation is the sum of the elements down the leading diagonals of the matrices of the representation.

Kronig–Penney Model

An approximate one-dimensional method of obtaining features of electron propagation in a crystal by considering the crystal as having a periodic square-well type of variation of electric potential. The potential is assumed, in the limit, to be a periodic delta function of infinite height and zero width but finite area, and the method obtains discrete ranges of energy for the electrons.

Kubo Relations

Expressions used for calculating such transport parameters as electrical conductivity. The expressions involve correlation functions to relate a flux which arises in response to a perturbation applied to the system. Thus, for obtaining the electrical conductivity, the method involves time-correlations of the fluctuating components of the electrical current. The properties calculated are inherent in the unperturbed system and application of the perturbation exposes the time-correlations.

Kuhn–Thomas–Reiche *f*-sum Rule

For an atom whose electrons are undergoing all possible transitions from (or to) one level denoted subscript 2,

$$\sum_{1} f_{12} + \sum_{3} f_{23} = Z$$

where subscripts 1 and 3 refer respectively to all levels (including the continuum) below and above the level denoted 2, the terms f are the oscillator strengths for each transition, and Z is the number of optical electrons.

Kummer's Transformation

If $s = \sum_{k=0}^{\infty} a_k$ and $S = \sum_{k=0}^{\infty} b_k$ are two convergent series such that

$L = \lim_{n \to \infty} (a_n/b_n) \neq 0$, then

$$s = LS + \sum_{k=0}^{\infty} \left(1 - L\frac{b_k}{a_k}\right) a_k.$$

Kundt's Constant
A constant used to measure the **Faraday Effect**. It is similar to **Verdet's Constant**, but is for rotation per unit magnetization in place of rotation per unit magnetic field.

Kundt's Rule
The refractive index of a medium on the shorter-wavelength side of an absorption band is abnormally low and on the longer-wavelength side abnormally high. The phenomenon of anomalous dispersion is due to this effect.

Kurie Plot (for Beta Decay)
In 1934 E. Fermi derived, for the energy distribution of beta particles emitted from radioactive isotopes, a formula of the form

$$\sqrt{\left[\frac{N(p)}{p^2 F(Z, p)}\right]} = C(E_{\max} - E)$$

where $N(p)$ is the number of particles having momentum between p and $p + dp$, $F(Z, p)$ is the correction factor for the coulombic interactions between the nuclear particles and the beta particles, E is the energy of the beta particles, which can be emitted with a maximum energy $E_{\max}$, C is a constant and Z is the atomic number of the product nucleus.

The distribution can be plotted in 'straight-line' form when the graph is called a Kurie plot or **Fermi plot**.

L

Ladenburg Benzene Formula
Ladenburg postulated a three-dimensional structure for benzene

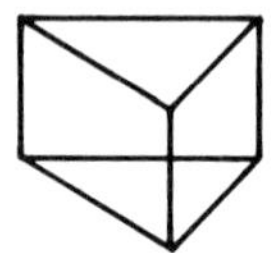

which was proved incorrect when the hydrogen atoms were found to be equivalent. *See also* **Kekulé Benzene Formula**

Lagrange's Differential Equation
An equation of the form

$$y = xf_1\left(\frac{\mathrm{d}y}{\mathrm{d}x}\right) + f_2\left(\frac{\mathrm{d}y}{\mathrm{d}x}\right)$$

(alternatively called **D'Alembert's Differential Equation**). It is solved by differentiating with respect to x and treating $\mathrm{d}y/\mathrm{d}x$ as the independent variable. The general solution is given parametrically in terms of $\mathrm{d}y/\mathrm{d}x$, whereas the singular solution is given by

$$y = xf_1(m) + f_2(m)$$

where m is the root of

$$f_1(m) - m = 0.$$

Lagrange's Equations of Motion
Consider a general physical system with n degrees of freedom expressed in generalized coordinates $q_1, q_2, \ldots q_n$.

Let $V = V(q_r, t)$ be the potential energy of the system, if conservative.

Let $T = T(q_r, \dot{q}_r, t)$ be the kinetic energy of the system

$\dot{q}_r = \mathrm{d}q_r/\mathrm{d}t$). Then the system can be described by the set of Lagrange's equations of state

$$\frac{\mathrm{d}}{\mathrm{d}t}\frac{\partial T}{\partial \dot{q}_i} - \frac{\partial T}{\partial q_i} = -\frac{\partial V}{\partial q_i} \quad (i = 1, 2, \ldots n)$$

Lagrange's Identity
If a_1, b_1, c_1, a_2, b_2 and c_2 are any numbers, then

$$(a_1 b_2 - a_2 b_1)^2 + (b_1 c_2 - b_2 c_1)^2 + (c_1 a_2 - c_2 a_1)^2$$
$$\equiv (a_1{}^2 + b_1{}^2 + c_1{}^2)(a_2{}^2 + b_2{}^2 + c_2{}^2) - (a_1 a_2 + b_1 b_2 + c_1 c_2)^2$$

Lagrange's Interpolation Formula
If $\mathrm{f}(x)$ is given for $x = x_0, x_1, x_2 \ldots x_n$, then the function

$$\mathrm{g}(x) = \mathrm{f}(x_0)\frac{(x - x_1)(x - x_2)\ldots(x - x_n)}{(x_0 - x_1)(x_0 - x_2)\ldots(x_0 - x_n)} + \ldots$$
$$+ \mathrm{f}(x_n)\frac{(x - x_0)(x - x_1)\ldots(x - x_{n-1})}{(x_n - x_0)(x_n - x_1)\ldots(x_n - x_{n-1})}$$

tends to $\mathrm{f}(x_0)$ at $x = x_0$, to $\mathrm{f}(x_1)$ at $x = x_1$ etc. It is not especially useful as, for interpolation, it is not usually necessary to take account of all the given values of x throughout the entire range; $\mathrm{g}(x)$ will usually be determined mainly by the adjacent tabulated values. However, most interpolation formulae are derived from Lagrange's formula.

Lagrange's Law (Optics) *See* **Helmholtz's Relation**

Lagrange's Mean Value Theorem
If a function $y = \mathrm{f}(x)$ is continuous in the closed interval $[a, b]$ and has a continuous derivative in this interval, then there exists at least one point c between a and b such that

$$\frac{\mathrm{f}(b) - \mathrm{f}(a)}{b - a} = \mathrm{f}'(c) \quad (a < c < b)$$

Lagrange's Multipliers
To find stationary points of $z = \mathrm{f}(x, y)$ subject to the condition $g(x, y) = 0$, three equations

$$\frac{\partial f}{\partial x} + \lambda \frac{\partial g}{\partial x} = 0$$

$$\frac{\partial f}{\partial y} + \lambda \frac{\partial g}{\partial y} = 0$$

$$g = 0$$

are solved simultaneously giving Lagrange's multiplier λ, and the stationary point x, y.

Lagrange's Theorem
If $y = z + x\phi(y)$,

$$\mathrm{f}(y) = \mathrm{f}(z) + x\phi(z)\mathrm{f}'(z) + \frac{x^2}{2!}\frac{\mathrm{d}}{\mathrm{d}z}[\{\phi(z)\}^2\,\mathrm{f}'(z)] + \ldots$$

$$+ \frac{x^n}{n!}\frac{\mathrm{d}^{n-1}}{\mathrm{d}z^{n-1}}[\{\phi(z)\}^n\,\mathrm{f}'(z)] + \ldots$$

Lagrange's Theorem in Group Theory
The order of a subgroup is a factor of the order of the group.

Lagrange's Theorem of Divisibility
If p is a prime, and r is any number less than $p-1$, the sum of the products of the numbers $1, 2, 3, \ldots p-1$ taken r together is divisible by p.

Lagrangian
The Lagrangian

$$L = T - V$$

sometimes called kinetic potential, is the difference between the kinetic and potential energies of a system expressed in generalized coordinates (cf. **Lagrange's Equations of Motion**). The function is normally used to express a system possessing a finite number of degrees of freedom. For a continuous system, for example a vibrating elastic solid, the Lagrangian is expressed as an integral over the coordinates x, y and z, which are continuous variables. For every value of x, y and z

there is a generalized coordinate $\eta(x, y, z, t)$ and the Lagrangian is given by

$$L = \iiint \mathscr{L} \, dx \, dy \, dz$$

where

$$\mathscr{L} = \mathscr{L}\left(\frac{d\eta}{dx}, \frac{d\eta}{dy}, \frac{d\eta}{dz}, \frac{d\eta}{dt}, x, y, z, t\right)$$

is called the **Lagrangian density function.**

Laguerre Polynomial
A solution of Laguerre's differential equation

$$x\frac{d^2y}{dx^2} + (1 - x)\frac{dy}{dx} + ny = 0$$

where n is an integer, in the form

$$y = L_n(x) = e^x \frac{d^n}{dx^n}(x^n e^{-x})$$

In addition, the associated (or generalized) Laguerre polynomial given by

$$y = L_n^k(x) = \frac{d^k}{dx^k} L_n(x)$$

is a solution of the equation

$$x\frac{d^2y}{dx^2} + (k + 1 - x)\frac{dy}{dx} + (n - k)y = 0$$

Lambert *See* **Appendix**

Lambert's Law of Absorption
If the thickness of an absorbing medium increases in arithmetic progression, the light transmitted decreases in geometrical progression.

$$I = I_0 \exp(-kx)$$

where I is the intensity of light after passage through a layer of thickness x and I_0 is the intensity of the incident light. This is also known as **Bouguer's law**.

Lambert's Law of Emission (Lambert's Cosine Law)
The intensity of the light emitted from a perfectly diffusing surface is proportional to the cosine of the angle between the normal to the surface and the direction of observation.

Lamb Shift (or Lamb–Rutherford Shift)
Spectroscopists found a separation between the main components of the H_α line in the hydrogen spectrum to be less than predicted by the theory of Dirac. This was shown by Lamb, using quantum electrodynamic theory, to be due to the $2S_{1/2}$ term (quantum numbers $\ell = 1, j = \frac{1}{2}$) being 0·2836 m^{-1} higher than the $2P_{1/2}$ term (quantum numbers $l = 0, j = \frac{1}{2}$), this difference being called the Lamb shift.

Lamé Curves
Curves of the form

$$(x/a)^{2/3} \pm (y/b)^{2/3} = 1.$$

Lamé's Elastic Constants
The elastic properties of an isotropic material are completely determined by two constants called Lamé's elastic constants and denoted by μ and λ. μ is the shear modulus or rigidity coefficient (also denoted G). λ is not a named modulus but is related to the bulk modulus, K, by

$$K = \lambda + \tfrac{2}{3}\mu$$

Lamé's Functions
Lamé's equation

$$\frac{d^2 z}{dx^2} = \{m(m+1)\mathscr{P}(x) + B\}z$$

where m is a positive integer and $\mathscr{P}(x)$ is the **Weierstrass Function**, has $2m + 1$ values of B for which the equation has a solution of one or the other of the four species of ellipsoidal harmonics. If, when such a solution is expanded in descending powers of the argument, the coefficient of the leading term is taken as unity, the function so obtained is termed a Lamé function of degree *m*, of the first kind and of the species corresponding to the ellipsoidal harmonic taken.

Lami's Theorem
If three forces act on a particle in equilibrium, each is proportional to the sine of the angle between the other two.

Landau Diamagnetism
This is the net non-vanishing magnetization which is produced antiparallel to an applied magnetic field as a result of coupling of the magnetic field with orbital electronic motion. For free electrons, the diamagnetic susceptibility has one third the magnitude of the paramagnetic **Pauli Spin Susceptibility**.

Landau Fluctuations
Experimental determination of the energy loss of a fast particle frequently involves measurement of the energy lost in a thin detector in which the particles suffer many ionizing collisions but do not lose much of their initial energy. There are marked fluctuations in the observed rate of energy loss. This variation may be attributed to the **Poisson Distribution** of ionizing events and to the wide range of energies which can be lost by fast particles in an inelastic collision.

Landau Levels
When a magnetic field is applied to a metal at low temperatures, the electrons are forced to coalesce into quantized levels of energy $(n+\gamma)\hbar\omega_c$. n is an integer, γ takes the value 0·5 if free electrons are assumed and ω_c is the cyclotron frequency corresponding to the magnetic field. These energy levels are called Landau levels.

Lande Splitting Factor
Lande, partly empirically, derived a formula for g, the fraction by which the magnetic moment of an atom must be multiplied to allow for spin, in terms of the three quantum vectors: J the total momentum vector, L the azimuthal quantum vector and S the moment of electron spin:

$$g = 1 + \frac{J(J+1)+S(S+1)-L(L+1)}{2J(J+1)}$$

Landolt Fringe
If a brilliant source of light is viewed through two Nicol prisms orientated with their principal axes perpendicular to one another, the field is not quite dark when exact adjustment is made. The darkened field is crossed by a black fringe, which changes position if the prisms are rotated slightly. Lippich showed that this fringe was due to the fact that the directions of vibrations in various parts of the field were not parallel.

Lane's Law
If a star contracts, its internal temperature must rise. The law is independent of the source of energy but assumes that the material of the star behaves as a perfect gas.

Langevin Diamagnetism Equation
The diamagnetic susceptibility per gram-molecule for a diamagnetic substance is given by

$$\chi_m = -\frac{\mu_0 Z e^2 N_A}{6m}\langle r^2 \rangle$$

where N_A is Avogadro's number, Z the atomic number and $\langle r^2 \rangle$ the mean square distance of the electrons from the nucleus, assuming spherical symmetry.

Langevin Function
Defined by $L(x) = \coth x - (1/x)$, the function is used for the calculation of magnetization in paramagnetic substances and the electric polarization in substances containing molecules with a permanent electric dipole moment. For the former, magnetization per mole is given by

$$M = N_A \mu L(x) \quad \text{where} \quad x = \frac{\mu B}{k_B T}$$

μ is the magnetic moment per molecule, B is the magnetic induction and N_A is Avogadro's number. Electric polarization is obtained by replacing μ by p, the electric dipole moment and B by E, the electric field, in the above equations. The function applies strictly to gases, where the molecules are sufficiently far apart for their mutual interactions to be negligible. (*See also* **Brillouin Function**.)

Langley *See* **Appendix**

Langmuir Adsorption Isotherm
From a consideration of the number of gas molecules striking and leaving the surface of the adsorbent, and assuming the adsorbed layer to be monomolecular, Langmuir showed that the fraction of surface covered is

$$\frac{V}{V_{\mathrm{m}}} = \frac{kp}{1+kp}$$

where V is the volume of gas adsorbed, V_{m} is the volume of gas necessary to cover the whole surface, k is a constant for a system and p is the pressure of the gas.

Langmuir–Child Law *See* **Child's Law**

Laplace's Coefficients
Laplace's equation $\nabla^2 V = 0$ becomes, in spherical coordinates:

$$r\frac{\partial^2}{\partial r^2}(rV) + \frac{1}{\sin\theta}\frac{\partial}{\partial\theta}\left(\sin\theta\frac{\partial V}{\partial\theta}\right) + \frac{1}{\sin^2\theta}\frac{\partial^2 V}{\partial\phi^2} = 0$$

Particular solutions of this equation are

$$r^m \mathrm{Y}_m(\cos\theta, \phi) \quad \text{and} \quad \frac{1}{r^{m+1}}\mathrm{Y}_m(\cos\theta, \phi)$$

where m is a positive integer and

$$\mathrm{Y}_m(\cos\theta, \phi) = A_0 P_m(\cos\theta) + \sum_{n=1}^{n=m}\{A_n \cos n\phi\ \mathrm{P}_m^n(\cos\theta) + B_n \sin n\phi\ \mathrm{P}_m^n(\cos\theta)\}$$

where

$$\mathrm{P}_m^n(\cos\theta) = \sin^n\theta\,\frac{\mathrm{d}^n.\mathrm{P}_m(\cos\theta)}{\mathrm{d}(\cos\theta)^n}$$

is a Laplace coefficient or a surface spherical harmonic of the mth degree and $\mathrm{P}_m(\cos\theta)$ is a **Legendre Coefficient**.

Laplace's Equation
If, within a closed surface, there exists no charge, the potential on the surface is given by

$$\nabla^2 V = \frac{\partial^2 V}{\partial x^2} + \frac{\partial^2 V}{\partial y^2} + \frac{\partial^2 V}{\partial z^2} = 0$$

(*see* **Poisson's Equation**).

Laplace's Equation (for the Velocity of Sound)

$$v = \sqrt{\frac{\gamma p}{\rho}}$$

where v is the velocity of sound in a gas, γ is the ratio of the specific heats of the gas, p is the pressure and ρ the density.

Laplace's Integrals

$$\mathrm{P}_m(x) = \frac{1}{\pi} \int_0^{\pi} \{x + \sqrt{(x^2-1)} \cos \phi\}^m \, \mathrm{d}\phi$$

$$\mathrm{Q}_m(x) = \frac{1}{\pi} \int_0^{\infty} \frac{\mathrm{d}\phi}{\{x + \sqrt{(x^2-1)} \cosh \phi\}^{m+1}}$$

$\mathrm{P}_m(x)$ and $\mathrm{Q}_m(x)$ are **Legendre's Coefficients.**

Laplace's Law *See* **Ampère's Law**

Laplace's Principle
An irrationality in a function cannot be removed by differentiation. Conversely, no irrationality which does not occur in an integrand can appear after integration.

Laplace's Transform
If $\mathrm{f}(x) = 0$ for $x < 0$, if

$$\int_0^{\infty} |\mathrm{f}(x)|^2 \, \mathrm{e}^{-2\alpha x} \, \mathrm{d}x$$

is finite for $\alpha > \beta$ and if

$$\mathrm{F}(p) = \int_0^{\infty} \mathrm{f}(x) \mathrm{e}^{-px} \, \mathrm{d}x,$$

then F(p) is called the Laplace Transform of f(x) and

$$\mathrm{f}(x) = \frac{1}{2\pi \mathrm{j}} \int_{-\mathrm{j}\infty + \alpha}^{+\mathrm{j}\infty + \alpha} \mathrm{F}(p) \mathrm{e}^{px} \, \mathrm{d}x \ (\alpha > \beta;\ x > 0)$$

Laplace–Young Equation
An equation relating the excess pressure p across a curved fluid surface, or surface separating two fluids, at a point with two principal radii of curvature R_1 and R_2 (maximum and minimum radii of curvature of the surface at the point).

$$p = \sigma\left(\frac{1}{R_1} + \frac{1}{R_2}\right)$$

where σ is the surface tension of the fluid at the interface.

Laplacian Operator
The Laplacian operator ∇^2 is defined by $\nabla^2 \equiv (\nabla . \nabla)$. For rectangular Cartesian coordinates:

$$\nabla^2 \equiv \frac{\partial^2}{\partial x^2} + \frac{\partial^2}{\partial y^2} + \frac{\partial^2}{\partial z^2}$$

Laporte's Rule
For dipole radiation, only transitions between terms of opposite parity are allowed: i.e. transitions between terms having either $\Sigma_i l_i$ even or having $\Sigma_i l_i$ odd are allowed, where $\Sigma_i l_i$ is the summation of the absolute value of the quantum number l for all the electrons.

Larmor Diamagnetism
An alternative name for Langevin diamagnetism. *See* **Langevin Diamagnetism Equation.**

Larmor Precession
The action of a magnetic field on the orbital motion of an electron was stated by Larmor. He showed that a uniform field does not affect the form and inclination of the orbit to the magnetic lines of force, or the frequency and velocity of the electron in its orbit. The orbit itself, however, rotates (precesses) in a uniform fashion about an axis

parallel to the field, and the frequency of precession is called the **Larmor frequency**.

Larmor's Formula
A single accelerating electron radiates energy at the rate of

$$\frac{e^2}{6\pi\varepsilon_0 c^3}\left(\frac{\mathrm{d}v}{\mathrm{d}t}\right)^2$$

where v is the velocity of the electron and the formula is in SI units.

Laue Equations
When diffraction of x-rays is produced by a three-dimensional array of atoms, the condition for scattering in phase can be broken down into three equations called the Laue equations, which must be satisfied simultaneously. If there are three crystal axes x, y and z, and the separations of the atoms along the three axes are a, b and c respectively, then the equations are

$$a(\cos\alpha - \cos\alpha_0) = h\lambda$$

$$b(\cos\beta - \cos\beta_0) = k\lambda$$

$$c(\cos\gamma - \cos\gamma_0) = l\lambda$$

Angles α_0, β_0 and γ_0 are the angles which the incident beam makes with the x, y and z axes respectively and α, β and γ are the corresponding angles for the diffracted beam. λ is the wavelength of the radiation and h, k and l are integers corresponding to the **Miller Indices** of the planes considered to be diffracting the x-rays. (*See also* **Bragg's Law**, which is another form of these equations.)

Laue Symmetry Groups *See* **Friedel's Law**

Laurent's Expansion
Suppose the point a is a regular point or a pole or an isolated essential singularity of $\mathrm{f}(z)$, but that $\mathrm{f}(z)$ is analytic within the annulus between two concentric circles whose centre is a and whose radii are r_1 and r_2. Then, within the region R

$$\mathrm{f}(z) = \sum_{n=-\infty}^{n=+\infty} C_n(z-a)^n$$

where

$$C_n = \frac{1}{2\pi j} \int_C \frac{f(z)\,dz}{(z-a)^{n+1}} \quad (n = 0, \pm 1, \pm 2, \ldots)$$

Laves Phases

Intermediate alloy phases having the general formula AB_2 and possessing one of three related crystals structures: $MgCu_2$ type (cubic), $MgZn_2$ type or $MgNi_2$ type (both hexagonal).

Lebesgue Integral

If $\{\phi_n(x)\}$ is a sequence of step functions bounded almost everywhere, and if this sequence $\{\phi_n(x)\}$ converges almost everywhere to $f(x)$, then by definition the Lebesgue integral is

$$\int_a^b f(x)\,dx \equiv \lim_{n \to \infty} \int_a^b \phi_n(x)\,dx$$

and $f(x)$ is called summable and integrable in the Lebesgue sense.

Leblanc Process (for the Manufacture of Sodium Carbonate)

This process in no longer in use in Britain, having given way to the ammonia–soda or **Solvay Process** and the electrolytic processes.

The process occurred in two parts. Common salt was heated with sulphuric acid. The sodium sulphate formed was ground and mixed with its own weight of chalk and half its weight of coal or coke, and fused in a rotating furnace. The mixture of carbonate and sulphide was extracted with water and the carbonate crystallized out.

Le Chatelier's Principle

If a change occurs in one of the conditions of a system in equilibrium, the system will adjust itself so as to annul, as far as possible, the effect of that change.

Lederer–Manasse Reaction

Phenol condenses with aliphatic and aromatic aldehydes in the *o*- and *p*-positions. The most important application of this reaction is the condensation of 40 per cent formaldehyde in the presence of dilute alkali or acid.

The main product is the *p*-isomer and the reaction is the primary step in the production of phenol–formaldehyde resins.

Leduc Effect *Also called the* **Righi–Leduc Effect**

Legendre Coefficients

The general solution of Legendre's equation

$$(1-x^2)\frac{d^2z}{dx^2}-2x\frac{dz}{dx}+m(m+1)z=0$$

where m is a positive integer, is

$$z = A\,P_m(x)+B\,Q_m(x).$$

$$P_m(x)=\frac{(2m-1)\,(2m-3)\,.\,.\,.\,1}{m!}\left\{x^m-\frac{m(m-1)}{2(2m-1)}x^{m-2}\right.$$

$$\left.+\frac{m(m-1)\,(m-2)\,(m-3)}{2.4.(2m-1)\,(2m-3)}x^{m-4}-\,.\,.\,.\right\}$$

is a finite sum finishing with the term which involves x if m is odd, or $x^0=1$ if m is even, and is called a Legendre coefficient or a surface zonal harmonic. $Q_m(x)$, a similar sum, which however has a logarithmic singularity at $x=0$, is called a surface zonal harmonic of the second kind.

Legendre's Elliptic Integrals

In the problem of the rectification of the ellipse, the two integrals

$$F(k,\phi)=\int_0^{\phi}\frac{d\phi}{\sqrt{(1-k^2\sin^2\phi)}}=u\ (0<k<1)$$

and

$$E(k,\phi)=\int_0^{\phi}\sqrt{(1-k^2\sin^2\phi)}\,d\phi\ \ (0<k<1)$$

arise. These are known as Legendre's forms of the elliptic integrals of the first and second kind.

The elliptic integral of the third kind may be written

$$\Pi(n,k,x)=\int_0^{\phi}\frac{1}{(1+n\sin^2\phi)\,\sqrt{(1-k^2\,sin^2\phi)}}d\phi\ (0<k<1)$$

where n is a constant.

ϕ is defined as the amplitude of u to the modulus k, i.e. $\phi=\text{am}\ u$, also $\sin\phi=\sin\ \text{am}\ u=\text{sn}\ u$, $\cos\phi=\text{cn}\ u$,

$\sqrt{(1-k^2\sin^2\phi)} = \text{dn}\,u$, and sn u, cn u and dn u are **Jacobi's Elliptic Functions.**

Leibnitz's Monads
In Leibnitz's philosophy, all things, spiritual and physical, were considered in terms of fundamental entities which he called monads. As far as their physical significance was concerned, the monads were 'the very atoms of nature' but they had neither mass, parts, extension nor figure. They existed solely as centres of force or of energy.

Leibnitz's Test (for Alternating Series)
If $a_n \to 0$ monotonically, then the series $\Sigma(-1)^n a_n$ is convergent.

Leibnitz's Theorem
If u and v are functions of x

$$\frac{\mathrm{d}^n(uv)}{\mathrm{d}x^n} = u\frac{\mathrm{d}^n v}{\mathrm{d}x^n} + \frac{n}{1!}\frac{\mathrm{d}u}{\mathrm{d}x}\cdot\frac{\mathrm{d}^{n-1}v}{\mathrm{d}x^{n-1}} + \frac{n(n-1)}{2!}\frac{\mathrm{d}^2 u}{\mathrm{d}x^2}\cdot\frac{\mathrm{d}^{n-2}v}{\mathrm{d}x^{n-2}} + \ldots + \frac{\mathrm{d}^n u}{\mathrm{d}x^n}v.$$

Leidenfrost's Phenomenon
If liquid is dropped on to a hot surface, there is a critical temperature above which wetting by the liquid does not occur. The liquid becomes insulated from the surface by a layer of vapour. As a result, water, for example, when dropped on to a hot plane, breaks into globules which move erratically across the surface.

Lennard-Jones Potential
The total potential energy $\Phi(r)$ of two atoms of an inert gas at separation r is

$$\Phi(r) = \frac{B}{r^{12}} - \frac{C}{r^6}$$

where B and C are empirical parameters. The r^{-6} term represents the long-range attractive force (*see* **Van der Waals–London Interaction**) and the r^{-12} term the short-range repulsive force.

Lenz's Law
If an e.m.f. is induced in a circuit due to a change in the flux of magnetic induction through the circuit, the sign of the e.m.f. is such

that any current flow is in a direction that would oppose the flux change.

Leuckart Reaction
Aldehydes and ketones are converted into the formyl derivatives of the corresponding amine by excess ammonium formate or formamide.

Lewis Acids
Reagents which act as electron-pair acceptors. They must have either an empty orbital or a potentially empty orbital in order to accept the electron pair. Reagents which act as electron-pair donors are called **Lewis bases.**

Lewis Equation *See* **Nernst–Lindemann Equation**

Lewis Number
A dimensionless characteristic number (Le) for flow phenomena involving both heat and mass transfer. It is given by

$$(Le) = \frac{\lambda}{\rho C D}$$

where λ is thermal conductivity, ρ density, C specific heat and D the coefficient of diffusion.

Lewis–Randall Rule
At a given temperature the fugacity of a constituent of a mixture of gases is proportional to its mole fraction. Up to 25 atmospheres this rule appears to hold, and is a great improvement on **Dalton's Law of Partial Pressures.**

L'Hôpital's Cubic *See* **Catalan's Trisectrix**

L'Hôpital's Rule
If $f(x+L)$ and $F(x+L)$ can be both developed in Taylor's series (*see* **Taylor's Series**), and if $f(x)=0=F(x)$, the limit of the value of $f(x)/F(x)$ is $f'(x)/F'(x)$ provided this has a definite value (zero, finite or infinite). If not, the limit of the value is that of the first ratio of the derivatives that is definite.

Liebermann's Nitroso-Reaction

When secondary amines react with nitrous acid, oily nitrosamines are formed

$$R_2NH + HO.NO \rightarrow R_2N.NO + H_2O$$

This procedure is used as a test for secondary amines. If a nitrosamine is warmed with a crystal of phenol and a small quantity of concentrated sulphuric acid, a green solution is formed which turns blue on the addition of an alkali.

Lie Groups

S. Lie was the first to undertake a systematic study of the construction of transformation groups from their infinitesimal elements. Once the infinitesimal transformations are known, the continuous Lie group of transformations can be generated by integration.

Liénard–Wiechert Potentials

The scalar and vector potentials ϕ and $\mathbf{A}$ respectively at a distance r from a moving charge q:

$$\phi(\mathbf{r}, t) = \frac{q}{c[r - \mathbf{r}.\mathbf{v}/c]_{\text{ret.}}}$$

$$\mathbf{A}(\mathbf{r}, t) = \frac{q\mathbf{v}}{c[r - \mathbf{r}.\mathbf{v}/c]_{\text{ret.}}}$$

where $\mathbf{r}$ and $\mathbf{v}$ inside the square brackets are the position and velocity at the retarded time $t' = (t - r/c)$.

Liesegang Rings

It was observed by R. E. Liesegang (1896) that, when a drop of concentrated silver nitrate was placed on a film of gelatine containing potassium dichromate, silver chromate was not precipitated as a continuous film but in a series of well-defined rings. Many other instances of rhythmic precipitates are known. These rings may be obtained in solution, i.e. in the absence of a gel, provided convection currents in the solution are prevented.

Linde's Rule

An expression for the increase in resistivity, per atomic-per-cent impurity, for substitutional impurities introduced into a monovalent metal:

$$\Delta\rho = a + bz^2$$

where a and b are constants for a given row of the periodic table and a given solvent metal, and z is the excess valency (i.e. the impurity has valency $z+1$).

Liouville's Theorem (for a Function)
If a function of a complex variable $f(z)$ has no singular point for z (finite or infinite), then $f(z)$ is a constant.

Liouville's Theorem (Statistical Mechanics)
All elements of equal volume in phase space have equal *a priori* probabilities.

Lipschitz Condition
If

$$|f(\zeta) - f(x)| \leqslant A|\zeta - x|^{\alpha}$$

for given x and all $|\zeta - x| < \delta$ where A and α are independent of ζ, $\alpha > 0$ and δ can be arbitrarily small, $f(\zeta)$ is said to satisfy a Lipschitz condition of order α at $\zeta = x$. If $f(\zeta)$ satisfies a Lipschitz condition it is continuous at $\zeta = x$, and if it satisfies one for all x in $a \leqslant x \leqslant b$, then it is continuous within this closed interval.

Lissajous Figure
When a point undergoes two periodic motions, which at any instant of time are at right angles to one another, the resultant movement of the point traces out a curve which is called a Lissajous figure.

Lobry de Bruyn–van Ekenstein Rearrangement
In the presence of dilute alkalis or amines, polyhydric aldehydes and, in particular, sugars, undergo rearrangement. Thus if glucose is heated in the presence of sodium hydroxide, an inactive solution is obtained from which D(+) glucose, D(+) mannose and D(−) fructose have been isolated. The same products are obtained if the starting material is D(−) fructose or D(+) mannose. Possibly the rearrangement takes place through the 1:2-enediol.

London Equation for Superconductors

The London equation replaces **Ohm's Law** in superconductors and is

$$\mathbf{j} = -\frac{ne^2}{m}\mathbf{A}$$

where **A** the vector potential, is defined by curl **A** = **B**; **B** is the magnetic induction, n the density of 'superconducting' electrons and **j** is the current.

Sometimes the equation

$$\frac{d\mathbf{j}}{dt} = \frac{ne^2}{m}\mathbf{E},$$

where **E** is the electric field, is also called a London equation.

London Penetration Depth

The depth of penetration of an external magnetic field into a superconductor. It is given by

$$\lambda_L = (m/ne^2\mu_0)^{1/2}$$

where n is the density of 'superconducting' electrons. If the thickness of a superconducting film is much less than λ_L, then it will not exhibit the **Meissner Effect**.

London's Equation (for Intermolecular Attraction)

All molecules possess energy even at 0 K (null point energy) and this fact can only be explained by assuming that, even at this temperature, the nuclei and electrons vibrate in some way with respect to each other. As an instantaneous picture, the molecules will show various arrangements of the nuclei and electrons, these arrangements giving rise to dipole moments. A summation of these dipole moments over all the molecules will give a zero resultant. The cohesive force between molecules was attributed by London to the transient dipoles induced, in molecules in-phase with themselves, by these temporary dipoles. The interaction energy has been calculated, to a first approximation, as

$$\Phi_D = -\frac{3}{4}\cdot\frac{h\nu_0\alpha^2}{r^6}$$

where ν_0 is the characteristic frequency of the molecules, α is equal to $e^2/4\pi\varepsilon_0 k$ where e is the electronic charge, k is the restoring force constant, and r is the equilibrium distance between the positive ends of the dipoles.

Lorentz Force

The force **F** acting on a moving charge q in magnetic and electric fields is given in SI by

$$\mathbf{F} = q\left(\mathbf{E} + \mathbf{v} \times \mathbf{B}\right)$$

where **v** is the velocity of the charge, **E** the electric field and **B** the magnetic induction. Often only the magnetic term in the force is referred to as the Lorentz force. The force can then be expressed in the non-vectorial form

$$F = qvB \sin\theta.$$

θ is the angle between the direction of motion of the charge and the direction of the magnetic field, and the force is perpendicular to both the direction of motion and the field.

Lorentz Gauge

For an electromagnetic field, knowledge of the electric and magnetic fields is not sufficient to determine uniquely the scalar potential ϕ and the vector potential **A** of the field. Hence an auxiliary condition is chosen (the gauge) giving the equations a symmetrical form. The Lorentz gauge is the condition (in SI units):

$$\nabla.\mathbf{A} + \frac{1}{c^2}\frac{\partial\phi}{\partial t} = 0$$

Lorentzian Line Shape

This is the dependence of amplitude on frequency for a damped harmonic oscillator and can be represented by the expression

$$A(\omega) = \frac{(\frac{1}{2}\Gamma)^2}{(\omega_0 - \omega)^2 + (\frac{1}{2}\Gamma)^2}$$

where ω is angular frequency, ω_0 is the resonant frequency of the oscillator and Γ is the damping constant per unit mass for the oscillator. In nuclear physics the **Breit–Wigner Formula** is an expression of this type.

Lorentz–Lorenz Law

$$\frac{n^2-1}{n^2+2}=\frac{4\pi}{3}\Sigma\alpha$$

where n is the refractive index and $\Sigma\alpha$ is the sum of polarizabilities of all particles present in unit volume. α is defined by

$$p=\alpha E$$

where p is the electric dipole induced by an electric field E. *See also* **Clausius–Mosotti Equation.**

Lorentz Polarization Factor
In the scattering of x-rays by the atoms of a single crystal, the Lorentz polarization factor LP is given by

$$LP=\frac{1+\cos^2 2\theta}{2\sin 2\theta}$$

where θ is the Bragg angle. Tables of this factor versus θ are available.

A similar Lorentz polarization factor is available tabulated as a function of θ for scattering from a powder sample, as in the **Debye–Scherrer Method**. Here

$$LP=\frac{1+\cos^2 2\theta}{\sin^2\theta\cos\theta}$$

Lorentz Reciprocity Theorem
A theorem which relates the electric and magnetic fields generated by two sources of electromagnetic waves. The theorem has particular application with regard to antennas which, if constructed from linear isotropic material, have identical receiving and transmitting patterns.

Lorentz's Theory of Light Sources
When Maxwell showed that light was an electromagnetic wave, it was necessary to identify some electrical oscillator of atomic proportions to account for these waves.

Lorentz considered that electrons which are normally held in an equilibrium position are capable of vibration when displaced under the influence of a restoring force. The bound electron was supposed to be the source of the electromagnetic waves of light when executing damped vibrations. It was also capable of absorbing light when the frequencies of the vibration and of the light were in agreement.

Lorentz Transformation
The relation connecting the distance and time intervals between two Lorentz frames. In these, the coordinates of an event are x, y, z, t and x', y', z', t' and the transformation gives

$$x^2 + y^2 + z^2 - c^2t^2 = x'^2 + y'^2 + z'^2 - c^2t'^2.$$

The transformation can be represented by a 4 by 4 matrix and all transformations of this type make up what is called a general Lorentz group.

Lorenz Number
From **Wiedemann, Franz and Lorenz's Law**, the Lorenz number, L, is given by

$$L = \frac{\lambda}{\sigma T}$$

where λ is the thermal conductivity and σ the electrical conductivity. For metals the number is $k_B{}^2\pi^2/3e^2$but in semiconductors it depends on the **Fermi Level** and on the type of carrier present.

Loschmidt Number *See* **Appendix**

Lossen Rearrangement
This rearrangement is related to the **Hofmann** and **Curtius Rearrangements**. Hydroxamine acids will rearrange to give isocyanates.

Love Waves *See* **Rayleigh Waves**

Lucas' Reagent
A mixture of concentrated hydrochloric acid and zinc chloride very effective as a reagent for the conversion of alcohols to alkyl chlorides.

Lüder's Bands
In the plastic flow of crystals, surface markings which appear during yield failure. These bands appear at obvious places of stress concentration, such as the shoulder of a specimen, when the upper yield

point is reached. The boundaries of these surfaces divide the over-strained parts of the material from the parts which have not yielded.

Ludwig–Soret Effect Also called the **Soret Effect**

Lunge Scale *See* **Baumé Scale**

Lyddane–Sachs–Teller Relation
In an infinite ionic crystal, electromagnetic waves do not propagate in a frequency range $\omega_T < \omega < \omega_L$ where ω is the frequency of the electric field. ω_L is also the longitudinal optical phonon frequency for the phonon wave-vector **k** tending to zero and ω_T the transverse optical phonon frequency for a large wave vector. Then, if $\varepsilon_r(0)$ is the static dielectric constant of the crystal, and $\varepsilon_r(\infty)$ the dielectric constant at a frequency at which electronic polarizability is effective but not ionic polarizability, the Lyddane–Sachs–Teller relation is

$$\frac{\varepsilon_r(0)}{\varepsilon_r(\infty)} = \frac{\omega_L{}^2}{\omega_T{}^2}.$$

Lyman Series *See* **Balmer Series**

M

MacCullagh's Formula

If X is the origin at a point of a distribution of mass or charge and P is an external point distant r from X, then the potential at P due to the distribution is given by

$$\Phi = \gamma\frac{M}{r} + \frac{\gamma}{2r^3}(A + B + C - 3I) + \mathrm{O}\left(\frac{1}{r^4}\right)$$

where γ is the gravitational or electrostatic constant. M is the total mass or charge, I is the moment of inertia about XP and A, B and C are the principal moments of inertia about X.

Mach Effect

If the eye is presented with an illuminated field containing a discontinuity in the luminance which is both sharp and of high contrast, then a lightening of the field is observed adjacent to the light side and a darkening adjacent to the dark side. This effect gives rise to the **Craik–O'Brien illusion** (see figure) in which areas of similar luminance have different perceived brightnesses.

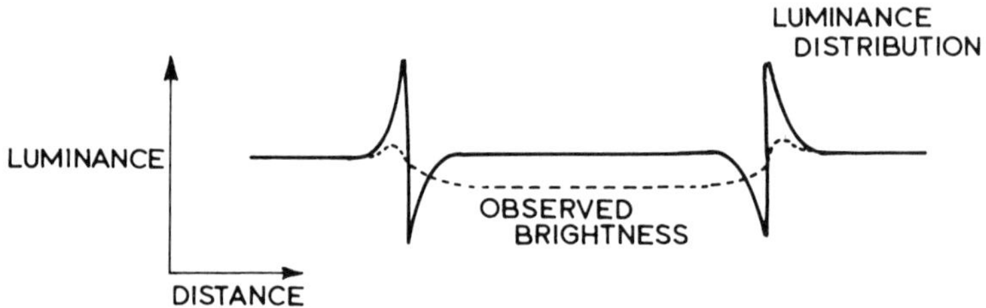

Mach Effect

Mach Number

The ratio of a velocity v to the velocity of sound in air, v_s, is termed the Mach number, M.

If $M = v/v_s < 1$, then v is said to be subsonic; if $v/v_s > 1$, v is supersonic.

The line or surface dividing the regions of subsonic and supersonic flow is called the **Mach line**.

Mach's Principle

The properties of space have no independent existence but depend on the mass content and distribution within it. Thus, acceleration dealt with in Newtonian mechanics can only have meaning if it is an acceleration with respect to the stars or with respect to something equally well-defined.

Maclaurin–Cauchy Test

If $f(x) > 0$ and $f(x) \to 0$ steadily as $x \to \infty$, then $\int_1^\infty f(x)\,dx$ and $\sum_{n=1}^{\infty} f(x)$ converge or diverge together.

Maclaurin's Theorem

A function of x may be expanded as a polynomial in x and derivatives of $f(x)$ when $x = 0$.

$$f(x) = f(0) + f'(0)\frac{x}{1!} + f''(0)\frac{x^2}{2!} + \ldots + f^n(0)\frac{x^n}{n!} + R_n$$

$$R_n = f^{n+1}(\theta x)\frac{x^{n+1}}{(n+1)!}(1-\theta)^n \quad (0 < \theta < 1)$$

This is a special case of **Taylor's Series** for $h = 0$.

Maclaurin's Trisectrix

A curve of polar form $r = a \sec(\theta/3)$.

Madelung's Constant

If ionic crystals are considered as composed of positively and negatively charged ions, the separation of pairs of ions being ρ_{ij}, Madelung's constant for a particular structure can be defined by

$$\alpha = \sum_{ij} (\pm)\rho_{ij}^{-1}$$

Maggi–Righi–Leduc Effect

Change of thermal conductivity of a conductor in a magnetic field.

Magnus' Effect
When a rapidly spinning body is hung up in an air current moving at right angles to its axis, it is deflected at right angles to both the current and its axis and moves to the side where its peripheral motion is in the same direction as the current.

Majorana Effect
When light is passed through certain liquid or crystalline materials in a direction perpendicular to an applied magnetic field, the material becomes doubly refracting.

Majorana Force
An exchange force between nucleons in which charge and spin are exchanged (equivalent to exchange of position). If the nucleons are represented by wave-functions, then if the wave-functions are symmetric with respect to interchange of the space coordinates the Majorana force is attractive; if the wave-functions are antisymmetric, the Majorana force is repulsive. (*See also* **Bartlett, Heisenberg** *and* **Wigner Forces**.)

Makarov–Zemlianski–Prokin Method
Benzylidene chloride usually contains benzyl chloride and benzotrichloride and, consequently, the hydrolysis product is contaminated with benzyl alcohol and benzoic acid. If, however, the hydrolysis is carried out by means of boric acid, only benzylidene chloride is hydrolysed, the other two being unaffected.

Malaprade Reaction
A procedure for the selective oxidation of vicinal glycols and hydroxyketones. The compound to be oxidized is treated with aqueous periodic acid or, if the compounds are water-insoluble, periodic acid in dioxane.

Malter Effect
The occurrence of large values of the coefficient of secondary electron emission in metals possessing a non-conducting surface film, notably in aluminium whose surface has been oxidized and then coated with caesium oxide. The surface film has a high resistance and becomes positively charged, and there is field emission of electrons from the metal.

Malus' Law
If a plane polarized beam of light is allowed to fall on a polarizer, the intensity of the transmitted beam is proportional to the square of the cosine of the angle between the plane of polarization of the incident light and the plane of polarization that would be required for total transmission of the beam.

Malus' Theorem
A system of rays normal to a wave front remains normal to the wave front after any number of refractions and reflections.

Mandel'shtam Effect
An alternative description of **Brillouin Scattering**.

Manly–Rowe Relations
These are equations used in the analysis of microwave parametric amplifiers but applicable to any system involving coupled bosons. If ϕ_1, ϕ_2 and ϕ_3 are photon fluxes at frequencies ω_1, ω_2 and ω_3 crossing any plane perpendicular to the direction of propagation in a non-absorbing, non-linear dielectric medium, then

$$\phi_1 + \phi_3 = \text{constant}$$
$$\phi_2 + \phi_3 = \text{constant.}$$

If ω_1 and ω_2 refer to inputs and ω_3 to output, then in a sum process if a photon is generated at frequency ω_3, single photons vanish at both ω_1 and ω_2.

Mannich Reaction
Aldehydes and ketones that have an α-hydroxyl react with formaldehyde and a secondary amine under weakly acidic conditions to give a dialkyl aminomethyl derivative of a carbonyl compound.

Mannich-type reactions are general for other compounds which lose a proton, such as acetylenes, nitroalkyls, phenols, thiophenes and pyrroles.

Marangoni Effect
The stabilizing effect observed in soap films such that, if an element of the film is stretched, there will be an increased force in this region tending to restore the film to its equilibrium configuration. This effect

arises because there is a decreased concentration of ions in the stretched region and, therefore, the surface tension in this area is increased because the surface tension of water is higher than that of soap solution.

Markoff Process
A probability process in which development of the process depends entirely on its state at any one time, and not on how that state arose. Every purely random process is a Markoff process.

Markownikoff's Replacement Rule
In the case of an unsymmetrical olefin, hydrogen halides can be added in two ways. Markownikoff's Rule states that the halogen is affixed to the carbon atom carrying the smaller number of hydrogen atoms. It has been shown, however, that normal addition can frequently be reversed in the presence of peroxy compounds.

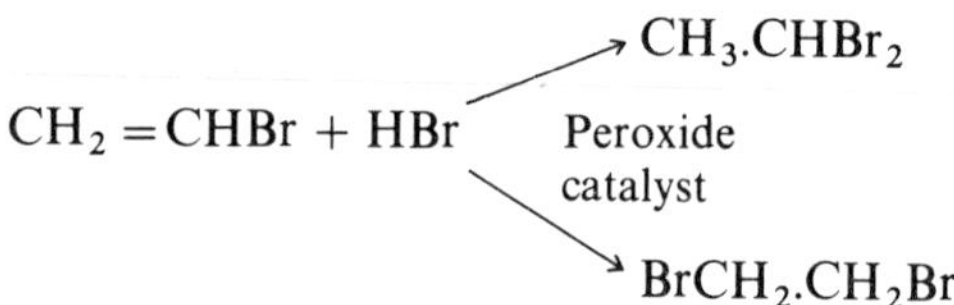

Marriotte's Law *See* **Boyle's Law**

Marsh Test for Arsenic
The substance to be tested is treated with arsenic-free zinc and hydrochloric acid to convert any arsenic present to arsine. The arsine is dried with $CaCl_2$ and freed from H_2S by lead acetate and is then decomposed by heat to give a characteristic mirror. Antimony gives a similar stain, which may be distinguished from the arsenic stain because it is not soluble in sodium hypochlorite solution.

Marx Effect
The reduction in the energy of photoemission by the simultaneous incidence of radiation of lower frequency than that producing the emission.

Mascheroni's Constant *See* **Euler's Constant**

Massieu Function
A thermodynamic function J given by

$$\mathrm{J} = -(U/T) + S$$

where U is internal energy and S entropy. It is equal to minus the quotient of the **Helmholtz Function** and the temperature T.

Mathieu's Function
Mathieu's functions are solutions of the Mathieu equation:

$$\frac{\mathrm{d}^2\chi}{\mathrm{d}\phi^2} + (b - h^2 \cos^2 \phi)\chi = 0$$

which occurs in problems involving systems with elliptical symmetry.
Depending on the value of b, the functions are periodic, stable or unstable.

Matteuci Effect
The appearance of an electric potential difference between the ends of a ferromagnetic specimen when twisted in a magnetic field. (*See also* **Barnett Effect.**)

Matthias' Rules
Empirical rules for the transition temperatures of superconducting metals and alloys:

I. The transition temperature of an element depends regularly on its position in the periodic table, but the transition metals show entirely different dependence on their valency to that exhibited by metals with completed d-shells.

II. For alloys between transition metals, or between non-transition metals, the transition temperature of the alloys is a smooth function of the mean valency v, weighted according to the atomic fractions of the constituent metals.

III. For elements and alloys with closed d-shells, the transition temperature rises from 0 K at $v = 2$ to a maximum at $v = 5$.

IV. In the transition series, the transition temperature has a maximum at $v = 5$ and $v = 7$ with a sharp minimum at $v = 6$.

Matthiessen's Rule
If the concentration of impurity atoms in a metal is small, the resistivity caused by scattering of electron waves by impurity atoms is independent of temperature. This resistivity is additive to that caused by the thermal motion of the periodic lattice of the metal, this latter resistivity being a function of temperature.

Matthiessen's Standard
A measure of the conductivity of copper expressed as a percentage of 0·1539 (if hard-drawn) or 0·1508 (if annealed) ohm for a specimen of length 1 metre and of weight 1 gram.

Maupertuis' Principle
The motion of a mechanical system is entirely determined by the principle of least action in which the integral $S = \int_{t_2}^{t_2} L(q_r, \dot{q}_r, t)\mathrm{d}t$ must be a minimum.

$L = L(q_r, \dot{q}_r, t)$ is the **Lagrangian**, q_r is a generalized coordinate and $\dot{q}_r = \mathrm{d}q/\mathrm{d}t$. The principle is analogous to **Fermat's Principle** in optics.

Maxwell *See* **Appendix**

Maxwell–Boltzmann Distribution Law
In a non-quantized system of N particles of mass m, the number of particles with velocities lying between $\mathbf{v}$ and $\mathbf{v}+\mathrm{d}\mathbf{v}$ is

$$\mathrm{d}N = 4\pi N\left(\frac{m}{2\pi k_{\mathrm{B}}T}\right)^{3/2}\mathbf{v}^2\exp\left(-\frac{m\mathbf{v}^2}{2k_{\mathrm{B}}T}\right)\mathrm{d}\mathbf{v}$$

Maxwell–Boltzmann Statistics
A method of studying the probabilities of the states of a system of non-quantized particles, defined by the average values of the position, velocity or energy coordinates, in a small but finite volume of the system.

Maxwell Effect (Flow Birefringence)
If a liquid is viscous, consists of anisotropic molecules, and flows in such a way that a shearing velocity gradient occurs, then it exhibits optical birefringence. If the liquid has velocity v_x (i.e. velocity in the x direction only), then for light travelling in the z direction the principal

planes of polarization are at 45° to the x and y axes, except for very high velocity gradients. The difference in refractive index is given by

$$\Delta n = \dot{c}\left(\frac{\partial v_x}{\partial y}\right)$$

where c is **Maxwell's constant**, which is a measure of birefringence.

Maxwell Equation (for Refractive Index)
For a non-absorbing non-magnetic material, the dielectric constant ε_r and refractive index n are related by the equation

$$\varepsilon_r = n^2.$$

Maxwell Primaries
The magenta, green and cyan (deep blue) colours forming the primary colours in Maxwell's three-colour additive synthesis.

Maxwell's Demons
Spontaneous changes in temperature and pressure such as to invalidate the Second Law of Thermodynamics could possibly be brought about by the intervention of intelligent beings: 'Maxwell's demons'. Thus, by means of a one-way valve, they might collect more molecules in one part of a system than in another, and thereby cause an increase of pressure. In fact experience shows that such 'demons' do not exist in closed systems in temperature equilibrium.

Maxwell's Field Equations
A set of four equations of classical electromagnetic theory. They relate certain vector quantities in a space which experiences changes in electric and magnetic fields. If **H** is the magnetic field, **B** the magnetic induction, **E** the electric field, **D** the electric displacement, ρ the electric charge density and j the conduction current density, then Maxwell's equations in SI units are:

$$\text{curl}\,\mathbf{H} = \frac{\partial \mathbf{D}}{\partial t} + \mathrm{j} \qquad \text{div}\,\mathbf{B} = 0$$

$$\text{curl}\,\mathbf{E} = -\frac{\partial \mathbf{B}}{\partial t} \qquad \text{div}\,\mathbf{D} = \rho$$

Maxwell's Rule

Every part of an electric circuit is acted upon by a force which tends to move it in such a direction as to enclose the maximum amount of magnetic flux.

Maxwell's Theorem

In an elastic structure, if a load applied at one point produces a given deflection at another point, then the same load applied at the point where the deflection was first measured will produce the same deflection as before at the point where the load was first applied.

Maxwell's Theory of Light

Clerk Maxwell showed in 1860 that the propagation of light could be regarded as an electromagnetic phenomenon, the wave consisting of an advance of coupled electric and magnetic forces. If an electric field is varied periodically, a periodically varying magnetic field is obtained which, in turn, generates a varying electric field; thus the disturbance is passed on in the form of a wave.

Maxwell's Thermodynamic Equations

For a thermodynamic system,

$$\left(\frac{\partial p}{\partial T}\right)_V = \left(\frac{\partial S}{\partial V}\right)_T \qquad \left(\frac{\partial V}{\partial T}\right)_p = -\left(\frac{\partial S}{\partial p}\right)_T$$

$$\left(\frac{\partial p}{\partial S}\right)_V = -\left(\frac{\partial T}{\partial V}\right)_S \qquad \left(\frac{\partial V}{\partial S}\right)_p = \left(\frac{\partial T}{\partial p}\right)_S$$

where p is the pressure, V the volume, T the absolute temperature and S the entropy.

Maxwell Stress Tensor

Maxwell considered the force between electric charges to be transmitted in a continuous way by means of tubes of force, which are considered as being in a state of tension. Hence, if a system in electrostatic equilibrium is divided into two parts by surface S, the force exerted on one part by the other must pass through the surface S. Maxwell showed this force to be given by

$$F = \iint_S T_n \, \mathrm{d}S$$

where T_n is the component matrix of the Maxwell stress tensor. In SI units the tensor is of the form:

$$T \equiv \begin{bmatrix} \varepsilon_r E_x{}^2 - \dfrac{\mathbf{E}^2}{2}\left(\varepsilon_r - \dfrac{d\varepsilon_r}{d\sigma}\sigma\right) & \varepsilon_r E_x E_y & \varepsilon_r E_x E_z \\ \varepsilon_r E_x E_y & \varepsilon_r E_y{}^2 - \dfrac{\mathbf{E}^2}{2}\left(\varepsilon_r - \dfrac{d\varepsilon_r}{d\sigma}\sigma\right) & \varepsilon_r E_y E_z \\ \varepsilon_r E_x E_z & \varepsilon_r E_y E_z & \varepsilon_r E_z{}^2 - \dfrac{\mathbf{E}^2}{2}\left(\varepsilon_r - \dfrac{d\varepsilon_r}{d\sigma}\sigma\right) \end{bmatrix}$$

where $\mathbf{E}$ is the electric field strength, E_x, E_y and E_z its x, y and z components, ε_r the dielectric constant and σ the charge density.

Mayer *See* **Appendix**

McLeod *See* **Appendix**

Meerwein–Ponndorf–Verley Reduction

When a carbonyl compound is heated with aluminium propoxide in isopropanol, the propoxide is oxidized to acetone and the carbonyl group is reduced to the alcohol. The acetone is removed from the equilibrium mixture by slow distillation. This reaction is specific for the carbonyl group and may be used in compounds containing other reducible substituents.

Meerwein Reaction

Diazonium halides in the presence of cuprous chloride add to a carbon–carbon double bond. The reaction works well for α, β-unsaturated carbonyl compounds and nitriles. It is usually carried out in acetone solution with catalytic quantities of cuprous chloride (e.g. **Sandmeyer Reaction**).

Mehler's Integrals

$$\mathrm{P}_m(\cos\theta) = \frac{2}{\pi}\int_0^{\theta} \frac{\cos(m+\frac{1}{2})\phi}{\sqrt{\{2(\cos\phi - \cos\theta)\}}}\,d\phi \quad 0 < \theta < \pi$$

for all values of m. If m is a positive integer

$$\mathrm{P}_m(\cos\theta) = \frac{2}{\pi}\int_{\theta}^{\pi} \frac{\sin(m+\frac{1}{2})\phi}{\sqrt{\{2(\cos\theta - \cos\phi)\}}}\,d\phi$$

where P_m is the **Legendre Coefficient**.

Meissner Effect
A bulk superconductor in a weak magnetic field exhibits perfect diamagnetism. Thus, if a specimen in the normal state is cooled through the transition temperature for superconductivity, the magnetic flux originally present in the sample is expelled, except for slight penetration at the surface. *See also* **London Equation for Superconductivity** *and* **London Penetration Depth**.

Mellin Transforms
If f(x) is defined (for $0 \leqslant x \leqslant \infty$) by the condition that

$$\int_0^{\infty} |f(x)|^2 x^{-2s-1}\, dx$$

converges for $s > s_0$, and if $M(p) = \int_0^{\infty} f(x) x^{p-1}\, dx,$

then $M(p)$ is called the Mellin transform of f(x) and conversely

$$f(x) = \frac{1}{2\pi j} \int_{-j\infty + s}^{j\infty + s} M(p) x^{-p}\, dp \quad (s > s_0;\, x > 0)$$

Mendeleev's Periodic Table
In 1869, three years after the publication of **Newland's Law of Octaves**, Mendeleev presented a paper to the Russian Chemical Society 'On the correlation of the Properties and Atomic Weights of the Elements'.

His 'Table of Elements' arranged the then-known elements into five periods, leaving blank spaces where necessary to establish periodicity and predicting that these blanks would be found to correspond with elements as yet unknown.

Mendius Reaction
Alkyl cyanides are reduced by nascent hydrogen to give primary amines.

Menelaus' Theorem
If ABC is a triangle and PQR a straight line cutting BC produced, CA, and also AB, at P, Q and R respectively, then

$$\frac{BP}{CP} \cdot \frac{CQ}{QA} \cdot \frac{AR}{RB} = 1$$

Mercator's Projection

A projection in which a sphere's surface (e.g. the earth's surface) is projected on to a cylinder taken about the equator. This transverse Mercator projection becomes increasingly inaccurate with increasing latitude, but alternative projections (oblique Mercator's projections) are made, where the cylinder's axis is taken at an angle other than 90° to the equator.

Mersenne Numbers

Numbers M given by $M = 2^p - 1$, where p is a prime number. It was originally thought that all the numbers M would be prime numbers, but this is not so.

Mersenne's Law

When a string is set into vibration, the frequency of the fundamental varies *directly* as the square root of the tension and *inversely* both as the length and as the square root of the mass per unit length of the string.

Mertens' Theorem

If Σu_n and Σv_n are convergent with sums s and t respectively, and one of these series, say Σu_n, is absolutely convergent, then Σd_n is convergent and its sum is st, where

$$\Sigma d_n = u_1 v_n + u_2 v_{n-1} + \ . \ . \ . \ + u_n v_1$$

Meusnier's Theorem

The osculating plane of a curve C at a point P_1 on a surface is the plane containing the limiting circle through P_1 and through two distinct points P_2 and P_3 (with P_2 and P_3 approaching P_1) and also containing the tangent to C at P_1. Then Meusnier's theorem states that, at P_1, the curvature κ of C is given by

$$\kappa = \left| \frac{\kappa_N}{\cos \alpha} \right|$$

where α is the angle between the osculating plane and the normal-section plane through the tangent, and κ_N is the normal-section curvature.

Meyer Atomic Volume Curve

If the atomic volumes of the elements are plotted against their atomic numbers, a periodicity is revealed. The alkali metals are seen to be on the peaks, whilst the transition elements appear in the valleys.

Michaelis Reaction
Phosphonic esters can be prepared from the reaction of a sodium salt of an alkyl phosphonate with an alkyl halide. (*See also* **Arbusov Reaction**.)

Michael Reaction
This reaction is a generalization of the **Claisen Condensation**. Other active methylene compounds such as ethyl phenyl acetate, benzyl cyanide and ethyl cyanoacetate also add to α, β-unsaturated ketones, esters, nitriles and sulphones.

Mie–Grüneisen Equation of State
An equation of state with particular application at high pressures.

$$p - p_0 = \frac{\gamma}{V}(U - U_0)$$

where p_0 and U_0 are the pressure and internal energy at 0 K and are functions of the volume at 0 K and γ is the **Grüneisen Number**, which is a function of volume V only.

Miller–Bravais Indices *See* **Miller Indices**

Miller Effect
In a single-valve amplifier, the apparent capacitance between grid and anode is multiplied, because of feedback, by $(1 - A)$ where A is the complex gain of the stage. A is usually large and negative in sign, a positive voltage at the grid giving rise to a negative voltage at the anode. The effect also occurs between the collector and the base of a transistor amplifier.

Miller Indices
A system of notation to represent uniquely the planes of atoms or molecules in a crystalline solid. The intercepts of a plane on each of the crystal axes are found, in units corresponding to the respective lattice constants for each of these three axes. The smallest integers having the same ratio as the reciprocals of these integers are called the Miller indices. The Miller indices of various planes contained in a cube of side a are shown in the diagram below.

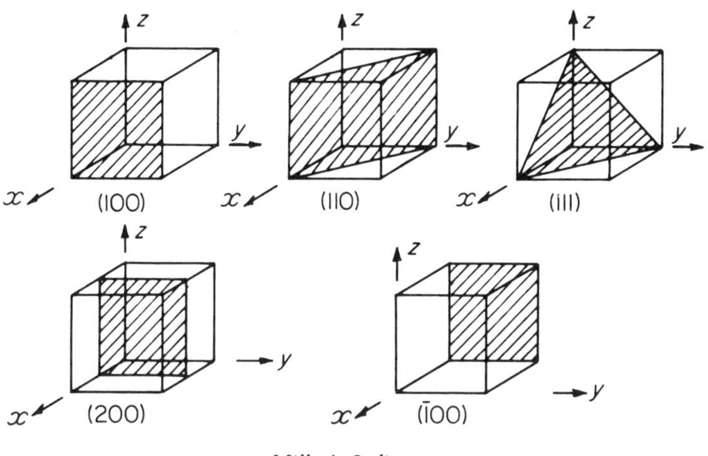

Miller's Indices

This system of notation has been modified to apply to the hexagonal crystal system and these indices are called **Miller–Bravais indices**. The method of calculation is the same as that given above, except that the intercepts are taken with respect to four axes. One of these axes is parallel to the major axis of a right hexagonal prism and the other three are in the basal plane at angles of 120° to each other. The simple Miller indices are written $(hk\ell)$, whereas the Miller–Bravais indices are written $(hki\ell)$. The first three indices which correspond to the intercepts in the basal plane are not independent, as $h + k + \ell = 0$. The symbols are therefore sometimes written $(hk \cdot \ell)$ or $(hk * \ell)$. For intercepts which occur on the axes in the negative directions, minus signs are inserted above the respective indices.

Miller's Law

In a crystal, the sine ratio of four faces in a zone is a commensurable number. (A zone consists of a set of planes whose intersecting edges are parallel.) If P_1, P_2, P_3 and P_4 are the four faces, and θ_{12} is the angle between P_1 and P_2 etc., then the sine ratio is given by

$$\frac{\dfrac{\sin\theta_{12}}{\sin\theta_{13}}}{\dfrac{\sin\theta_{42}}{\sin\theta_{43}}} = \frac{\dfrac{U_{12}}{U_{13}}}{\dfrac{U_{42}}{U_{43}}} = \frac{\dfrac{V_{12}}{V_{13}}}{\dfrac{V_{42}}{V_{43}}} = \frac{\dfrac{W_{12}}{W_{13}}}{\dfrac{W_{42}}{W_{43}}}$$

where $U_{12} = (k_1\ell_2 - \ell_1 k_2)$; $V_{12} = (\ell_1 h_2 - h_1\ell_2)$; $W_{12} = (h_1 k_2 - k_1 h_2)$ etc., $hk\ell$ being the **Miller Indices** of the faces.

Millington Reverberation Formula
In acoustics, a formula giving the reverberation time T in seconds as

$$0{\cdot}05\frac{V}{\sum_{r=1}^{n} S_r \ln(1-\alpha_r)}$$

where V is the volume in cubic feet and $\alpha_1, \alpha_2, \alpha_3, \ldots$ are the absorption coefficients of areas $S_1, S_2, S_3, \ldots$ (in square feet). It is also called the **Norris–Eyring reverberation formula.**

Millon's Base
Dihydroxomercuric-ammonium hydroxide

$$\left[\begin{matrix}\text{HO–Hg}\\ \text{HO–Hg}\end{matrix} > NH_2\right]OH$$

produced by the action of ammonia solution on yellow mercury oxide.

Millon's Reaction
When a protein is heated with a solution of mercuric nitrite and nitrate in nitrous and nitric acids (**Millon's reagent**) a red coloration or precipitate is formed.

Minkowski's Inequality

$$\left[\int_a^b |f_1 + f_2|^p \,dx\right]^{1/p} \leqslant \left[\int_a^b |f_1|^p \,dx\right]^{1/p} + \left[\int_a^b |f_2|^p \,dx\right]^{1/p} \qquad (1 \leqslant p < \infty)$$

The inequality implies a similar inequality for sums of convergent series

$$\left[\sum_j |a_j + b_j|^p\right]^{1/p} \leqslant \left[\sum_j |a_j|^p\right]^{1/p} + \left[\sum_j |b_j|^p\right]^{1/p}$$

See also **Hölder's Inequality**.

Minkowski Space
Four-dimensional space involving the x, y and z dimensions of real space and the dimension of time. The **Lorentz Transformation** can be considered as a rotation in Minkowski space.

Misterlich Law of Isomorphism
Substances which are similar in crystalline form and in chemical properties can usually be represented by similar formulae.

Möbius Band (Möbius Strip)
A one-sided surface formed by rotating one end of a strip of paper through 180° and then fastening the ends together.

Möbius Tetrads
Two tetrahedra arranged so that each vertex of either lies on a face of the other.

Mohr's Circle
A graphical method of transforming strains or stresses from one set of axes to another.

Mohr's Litre
The volume occupied by 1 kg of distilled water when weighed in air at 17·5°C against brass weights. It is approximately equal to 1·002 true litres.

Mohs' Hardness Scale
A scale of hardness of materials based on a scratching test wherein any material can be scratched by the material which follows it in the scale.

Moivre's Formula *See* **De Moivre's Theorem**

Molisch's Test for Carbohydrates
A sugar, when mixed with α-naphthol and treated with concentrated sulphuric acid, gives a violet coloration.

Møller Scattering
Electron scattering by another electron.

Monge's Theorem

By definition, centres of similitude are the points of intersection of lines joining the ends of parallel radii of two coplanar circles. Monge's theorem then states that the three outer centres of similitude of three circles taken in pairs are collinear. Any two of the inner centres of similitude are collinear with one of the outer centres.

Morse Equation

The potential energy of a diatomic molecule (Φ_r) according to P. M. Morse (1929) may be approximated by

$$\Phi_r = D\{1 - \exp[-a(r - r_e)]\}^2$$

where D is the dissociation energy, r is the distance apart of the atoms and r_e is the distance at equilibrium and a is a constant.

Moseley's Law

If x-rays are allowed to fall upon a substance, a series of characteristic x-ray lines is produced. Moseley showed that the frequency ν of the line belonging to any particular x-ray series is quantitatively related to the atomic number Z by the relationship

$$\sqrt{\nu} = a(Z - \sigma)$$

where a is the proportionality constant and σ is the same for all lines of a given series.

Mössbauer Effect

The Mössbauer effect is the recoil-less emission and resonant absorption of gamma radiation. When a nucleus decays from an excited state by emitting a gamma ray, the energy of the gamma ray is given by the energy level of the excited state less the energy imparted to the nucleus on recoil. The effect of the recoil of the nucleus is to reduce the gamma ray energy below the inter-level energy, such that a second nucleus in a lower state cannot be excited by gamma ray absorption. However, in 1958, Mössbauer found that, in certain systems (e.g. containing Fe^{57}), the recoil momentum is taken up by the whole crystal in which the nucleus is embedded. In this case, because of the much larger recoil mass, the energy of the emitted gamma ray is sufficient for it to excite a further nucleus. The uncertainty-spread in the frequency of the gamma ray is then as small as 1 part in 10^{13}. Hence the resonant absorption condition will be very sensitive to any relative velocity imparted between source and absorber, and a relative velocity of 1–10 mm s^{-1} can throw the absorber out of resonance. Thus using separate source and absorber, the effect can be used for energy resolution when, for instance, the energy levels of the source are split by the **Zeeman Effect**, and can also be used to repeat the Michelson–Morley experiment.

Mott Exciton *See* **Frenkel Exciton**

Mott–Hubbard Insulator *See* **Hubbard Model**

Mott Scattering

Coulombic-type scattering by nuclei in which transverse polarization of the scattered electrons is obtained.

Mott Transition

A type of sharp transition from non-metallic to metallic behaviour. The transition is a consequence of the interaction of the electrons, as this interaction can give rise to insulating behaviour even if the electrons do not fill an energy band, provided the band is narrow. If the band is widened, for example by decreasing the interatomic spacing, transition to a conducting state may occur. *See also* **Wilson Transition**.

N

Nabarro Circuit

A circuit used in disclinations for measuring the change in orientation of a lattice containing the disclination. It is constructed in an analogous way to the **Burgers' Circuit** for dislocations. Disclinations are imperfections involving warping or twisting of a lattice, and so can appear in thin periodic structures.

Napierian Logarithm

If $y = e^x$, where $e = \lim_{n \to \infty} [1 + (1/n)]^n = 2{\cdot}71828182\ldots$, then x is termed the Napierian logarithm of y.

Napier's Analogies *See* Delambre's Analogies

Napier's Rules

In the solution of right-angled spherical triangles: If a, b, c are the sides of the triangle and α, β, γ the angles opposite a, b, c respectively, c being the side opposite γ (the right angle), then the circular parts of the triangle are a, b, $\pi/2 - \alpha$, $\pi/2 - c$, $\pi/2 - \beta$. If these parts are supposed ranged inside a circle in the order in which they naturally occur with respect to the triangle, any one of these parts may be selected and called the middle part; the two parts next to it are the adjacent parts and the two remaining parts are called the opposite parts. Napier's rules are:

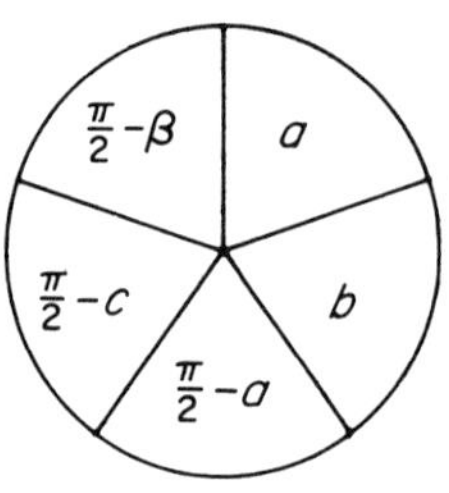

Napier's Rules

The sine of the middle part is equal to the product of the tangents of the adjaçent parts.

The sine of the middle part is equal to the product of the cosines of opposite parts.

Navier–Stokes Equation
Equation of motion for a viscous fluid. If the fluid is incompressible:

$$\frac{\partial \mathbf{v}}{\partial t} + \mathbf{v}.\,\text{grad}\,\mathbf{v} = \mathbf{F} - \frac{1}{\rho}\,\text{grad}\,p + \frac{\eta}{\rho}\nabla^2\mathbf{v}$$

where $\mathbf{v}$, p, $\mathbf{F}$ and ρ are as defined for **Euler's Equation** and η is the dynamic viscosity. For a compressible fluid, three equations are required of the type:

$$\frac{\partial v_x}{\partial t} = F_x - \frac{1}{\rho}\frac{\partial p}{\partial x} + \frac{1}{\rho}\frac{\partial}{\partial x}\left[\eta\left(2\frac{\partial v_x}{\partial x} - \frac{2}{3}\nabla\mathbf{v}\right)\right]$$
$$+\frac{1}{\rho}\frac{\partial}{\partial y}\left[\eta\left(\frac{\partial v_x}{\partial y} + \frac{\partial v_y}{\partial x}\right)\right]$$
$$+\frac{1}{\rho}\frac{\partial}{\partial z}\left[\eta\left(\frac{\partial v_z}{\partial x} + \frac{\partial v_x}{\partial z}\right)\right].$$

v_x, v_y and v_z are component velocities in the x, y and z directions in Cartesian coordinates and F_x is the component field of force in the x direction.

Néel Temperature
The temperature at which the spins of the neighbouring ions in an antiferromagnetic material cease to be ordered antiparallel as the temperature of the material is raised.

Néel Wall *See* **Bloch Wall**

Nef Reaction
If a primary or secondary nitro-compound is converted to the aciform by means of an alkali and then hydrolysed with 25 per cent sulphuric acid, aldehydes or ketones are produced with the evolution of nitrous oxide.

Neil's Parabola
A curve whose equation is

$$y = ax^{3/2}$$

Nelson–Riley Function
In the **Debye–Scherrer Method** of x-ray crystal analysis, the

percentage error in the calculation of the lattice parameter for given error in the measurement of diffraction angle approaches zero for a **Bragg Angle** of $\theta = 90°$. By plotting the calculated lattice parameter against the function

$$\frac{1}{2}\left(\frac{\cos^2\theta}{\sin\theta}+\frac{\cos^2\theta}{\theta}\right)$$

a linear extrapolation can be made to $\theta = 90°$. This function (available in tabulated form) was obtained by Nelson and Riley and, independently, by Taylor and Sinclair.

Neper *See* **Appendix**

Nernst Approximation Formula
Nernst was able to deduce the equilibrium constant of a gaseous reaction thermodynamically. From this relation, by a series of approximations which are now known to be inaccurate, he was able to obtain an equation which, though crude, can give useful information when no other data are available. The equation is

$$\log K' = \frac{-\Delta H}{4{\cdot}58\,T} + 1{\cdot}75\,\Sigma v_1 \log T + \Sigma v_1 C_1$$

where ΔH may be taken as the value of enthalpy measured at room temperature; Σv_1 is taken for the products minus reacting substances. For heterogeneous reactions, Σv_1 and $\Sigma v_1 C_1$ refer only to the gases, but ΔH to the whole reaction. C_1 is the conventional chemical constant, an empirical quantity, and v_1 is the number of reacting molecules of species 1.

Nernst Effect (or **Nernst–Ettingshausen Effect**)
If a temperature gradient is established in a conductor which is placed in a magnetic field that is orthogonal to the temperature gradient, then electrons tending to move from the hot to the cold end are initially deflected by the magnetic field. This produces a transverse electric field which prevents further deflection and which is given by

$$E_y = Q\,B_z\frac{\mathrm{d}T}{\mathrm{d}x}$$

where E_y is the Nernst electric field, Q the Nernst coefficient, B_z the magnetic flux density and dT/dx the longitudinal temperature gradient.

The above effect is sometimes referred to as the transverse Nernst–Ettingshausen effect. The variation of the **Seebeck Effect** in a magnetic field is then called the longitudinal Nernst–Ettingshausen effect.

Nernst Equation for E.M.F.

The potential $E_{\pm}$ of any reversible electrode immersed in a solution of a single ion of valency $Z_{\pm}$ is given by the equation

$$E_{\pm} = E_0 \mp \frac{RT}{Z_{\pm}F} \ln C_{\pm}$$

E_0 being the standard electrode potential, and $C_{\pm}$ the concentration of the ions.

Nernst Heat Theorem *See* **Nernst–Simon Statement**

Nernst–Lindemann Equation

An equation relating the specific heats C_p and C_V at constant pressure and volume respectively for a pure substance. It is used for calculating the variation of C_V with temperature.

$$C_p - C_V = A\, C_p^2\, T$$

where

$$A = \frac{V\gamma^2}{\kappa C_p^2}.$$

γ is the volume expansivity, κ the compressibility and A is a constant which can be obtained from the values of V, γ, κ and C_p for a single value of T. The equation is also called the **Lewis Equation**.

Nernst's Distribution Law

The partition of a substance between two solvents is governed by a law

$$\frac{C_1}{C_2^{1/n}} = \text{constant}$$

where C_1, C_2 are the concentrations of the substance in solvents 1, 2 and where the substance is present in monomolecular form A in solvent 1 and as polymeric molecules A_n in solvent 2.

Nernst–Simon Statement (of the Third Law of Thermodynamics)
The entropy change associated with any isothermal reversible process of a condensed system approaches zero as the temperature approaches zero. It is sometimes called the **Nernst heat theorem**.

Nernst Solution Pressure
In 1881, Nernst explained the existence of electrode potentials in terms of an equilibrium between a hypothetical solution pressure, which caused the atoms in the electrode material to pass into solution as positive ions, and the osmotic pressure of the ions in the solution, which caused the reverse process to occur. (*See also* **Nernst Equation for E.M.F.**)

Nernst–Thomson Rule
J. J. Thomson and W. Nernst in 1893 pointed out that, in a medium of low dielectric constant, the electrostatic attraction between the anions and cations of a dissolved electrolyte will be large, whilst in a solvent of high dielectric constant the attraction will be small. Hence solvents of high dielectric constant will favour dissociation, whilst solvents of low dielectric constant will have small dissociating influence.

Nernst Zero of Potential
An arbitrary zero of electrode potential, defined as that potential corresponding to the reversible equilibrium between hydrogen gas at 1 standard atmosphere pressure and hydrogen ions at unit activity.

Nesmeyanov Reaction
When mercury chloride is added to a solution of a diazonium chloride, a complex addition product is precipitated. In the Nesmeyanov reaction this complex is suspended in acetone and heated in the presence of copper powder to give an aryl mercuric chloride.

Neumann–Kopp Rule
The heat capacity of a gram-atom of a solid phase can be considered as the weighted sum of the heat capacities of the elements forming the

phase. The rule is applicable to compounds formed from elements in the solid or liquid states.

Neumann Problem
In potential theory, the determination of a function, harmonic in a region, when the boundary values of its normal derivatives are given (*see* **Dirichlet Problem**).

Neumann's Bessel Functions of the Second Kind *See* **Bessel Functions**

Neumann's Formula
This gives the mutual inductance between two circuits in the SI form

$$M_{12} = \frac{\mu}{4\pi}\oint_1 \oint_2 \frac{\mathrm{d}l_1 \, . \, \mathrm{d}l_2}{r}$$

where $\mathrm{d}l_1$ and $\mathrm{d}l_2$ are vector elements of currents in the two circuits, r is the distance between them and μ is the permeability. The integration is taken over all possible combinations of the scalar product of $\mathrm{d}l_1$ and $\mathrm{d}l_2$.

Neumann's Law
Whenever the flux of magnetic induction enclosed by a circuit is changing, there is an e.m.f. acting round the circuit, in addition to the e.m.f. of any batteries which may be in the circuit. The amount of this additional e.m.f. is equal to the rate of diminution of the flux of induction enclosed by the circuit.

Neumann's Principle
The symmetry elements of any physical property of a crystal must include the symmetry elements of the point group of the crystal.

Neumann's Series
An arbitrary function expanded in terms of **Bessel Functions**:

$$\phi(x) = \Sigma a_n \mathrm{J}_n(x)$$

is called Neumann's series.

Neumann's Triangle
The triangular equilibrium condition for three vectors $\mathbf{v}_1$, $\mathbf{v}_2$ and $\mathbf{v}_3$, i.e.

$$\mathbf{v}_1 + \mathbf{v}_2 + \mathbf{v}_3 = 0.$$

The condition applies, for example, to the surface tension forces for three liquids in equilibrium at a point.

Newland's Law of Octaves
If the elements were arranged in rising order of their atomic weights, starting from any one element, the eighth above it was of similar nature to the first. The rule, of course, breaks down after the first two short periods of the Periodic Table.

Newton *See* **Appendix**

Newton–Cotes Formula
A method of numerical integration in which the range of integration is divided into equally spaced intervals of width h. Then,

$$\int_{x_0}^{x_0+nh} y(x)\,\mathrm{d}x \approx A_0 y_0 + A_1 y_1 + \ldots A_n y_n = A.$$

However, for $n > 6$, it is usual to add m sums of $n' \leqslant 6$:

$$\int_{x_0}^{x_0+mn'h} y(x)\,\mathrm{d}x = \int_{x_0}^{x_0+n'h} y(x)\,\mathrm{d}x + \int_{x_0+n'h}^{x_0+2n'h} y(x)\,\mathrm{d}x + \ldots.$$

If $n' = 1$, this gives for each sub-interval

$$A_0 = A_1 = h/2,$$

so that when all m successive sub-intervals are included

$$A = h\left[\frac{y_0}{2} + y_1 + y_2 + \ldots y_{n-1} + \frac{y_n}{2}\right]$$

This is called the trapezoidal rule. $n' = 2$ and $n' = 6$ give, respectively, **Simpson's** and **Weddle's Rules**.

Newton–Gauss Interpolation Formula
A formula analogous to **Gregory's Interpolation Formula**.

$$f(x) = f(x_0) + \frac{x - x_0}{h}\Delta f(x_{1/2}) + \frac{(x - x_0)(x - x_0 - h)}{2!\,h^2}\Delta^2 f(x_0)$$

$$+ \frac{(x - x_0)(x - x_0 - h)(x - x_0 + h)}{3!\,h^3}\Delta^3 f(x_{1/2}) + \ldots$$

$$+ \frac{(x - x_0)(x - x_0 - h)(x - x_0 + h)\ldots\{x - x_0 - (n-1)h\}\{x - x_0 + (n-1)h\}}{(2n-1)!\,h^{2n-1}}\Delta^{2n-1} f(x_{1/2})$$

$$+ \frac{(x - x_0)(x - x_0 - h)(x - x_0 + h)\ldots\{x - x_0 - (n-1)h\}\{x - x_0 + (n-1)h\}\{x - x_0 - nh\}}{2n!\,h^{2n}}\Delta^{2n} f(x_0)$$

where

$$\Delta f(x_{1/2}) = f(x_0 + h) - f(x_0) \qquad \Delta f(x_{-1/2}) = f(x_0) - f(x_0 - h)$$

$$\Delta^2 f(x_0) = \Delta f(x_{1/2}) - \Delta f(x_{-1/2})$$

$$\Delta^3 f(x_{1/2}) = \Delta^2 f(x_0 + h) - \Delta^2 f(x_0) \text{ etc.}$$

Newtonian and Non-Newtonian Liquids
Liquids which show a linear relationship between velocity gradient and shearing stress are termed Newtonian liquids. Complex substances and high-molecular-weight and colloidal solutions do not exhibit this simple behaviour; the velocity gradient increases superlinearly with shearing stress, i.e. the viscosity of the liquid decreases with shearing stress. Such liquids are called non-Newtonian.

Newton–Raphson Formula
If it is required to find the root x of the equation $f(x) = 0$, then if x_n is a given approximation, an improved approximation is given by

$$x_{n+1} = x_n - \frac{f(x_n)}{f'(x_n)}$$

Newton's Approximation Method (for Roots of an Equation)
If $f(x) = 0$ has a root between γ and ζ and if $f(\gamma)$ and $f(\zeta)$ have opposite signs, an approximation of the root is given by

$$\alpha = \gamma - \frac{(\gamma - \zeta)\,f(\gamma)}{f(\gamma) - f(\zeta)}.$$

See also **Newton–Raphson Formula**.

Newton's Equation (for Conjugate Distances)
For a perfect lens system, if f and f' are the first and second focal lengths of the system, corresponding to space to the left and right of the system respectively, and x and x' are the distances of the object and image from the foci F and F′ respectively, as measured outwards from the lens system, then algebraically

$$xx' = ff'$$

Newton's Equation (for the Velocity of Sound)
This equation may be expressed in the form

$$v = \sqrt{\frac{p}{\rho}}$$

where v is the velocity of sound in a gas at pressure p and of density ρ, and is inaccurate as it assumes isothermal conditions.

Newton's Identities (for Sums of Powers of Roots)
Relations between the sums of powers of roots of the polynomial equation

$$\mathrm{f}(x) = a_0 x^n + a_1 x^{n-1} + \ldots a_r x^{n-1} + \ldots a_n = 0$$

$(a_0 \neq 0)$ and its coefficients. If $r_1, r_2, \ldots r_n$ are the roots and

$$s_k = \sum_{r=1}^{n} r^k \quad \text{and} \quad s_{-k} = \sum_{r=1}^{n} r^{-k}$$

where k is a positive number, then

for $k > n$: $a_0 s_k + a_1 s_{k-1} + \ldots a_r s_{k-r} + \ldots a_n s_{k-n} = 0$
for $0 < k \leqslant n$: $a_0 s_k + a_1 s_{k-1} + \ldots a_{k-1} s_1 + k a_k = 0$
for $a_n \neq 0$ and $k > n$: $a_n s_{-k} + a_{n-1} s_{1-k} + \ldots a_{n-r} s_{r-k} + \ldots a_0 s_{n-k} = 0$

and for $a_n \neq 0$ and $0 < k \leqslant n$:

$$a_n s_{-k} + a_{n-1} s_{1-k} + \ldots a_{n-k+1} s_{-1} + k a_{n-k} = 0$$

Newton's Interpolation Formula
If $\mathrm{f}(x)$ is given for $x = x_0, x_1, x_2, \ldots x_n$, then $\mathrm{f}(x)$ is given approximately by

$$\mathrm{f}(x) = \mathrm{f}(x_0) + (x - x_0)[x_0 x_1] + (x - x_0)(x - x_1)[x_0 x_1 x_2] + \ldots (x - x_0) \ldots (x - x_{n-1})[x_0 x_1 \ldots x_n]$$

where $[x_r x_{r+1}] = \dfrac{f(x_{r+1}) - f(x_r)}{x_{r+1} - x_r}$

and $[x\, x_0\, x_1 \ldots x_r] = \dfrac{[x_0\, x_1 \ldots x_r] - [x\, x_0\, x_1 \ldots x_{r-1}]}{x_r - x}$

Newton's Law of Cooling
The rate at which heat is lost by a heated body is proportional to the difference of temperature between the body and the surrounding medium when the temperature difference is small.

Newton's Law of Gravitation
A particle of mass m_1 attracts a particle of mass m_2 a distance d away with a force

$$F = G\frac{m_1 m_2}{d^2}$$

in the direction of the line joining the particles. If m_1 and m_2 are in kilograms, d is in metres and F in newtons, then

$$G = 6{\cdot}673 \times 10^{-11}\,\mathrm{N\,m^2\,kg^{-2}}.$$

Newton's Law of Resistance
At moderate velocities the resistance of a medium is proportional to the square of the velocity.

Newton's Laws of Motion
I. Every body continues in its state of rest or of uniform motion in a straight line except in so far as it may be compelled by impressed force to change that state.
II. Change of motion is proportional to the impressed force and takes place in the direction of the straight line in which the force acts.
III. To every action there is an equal and opposite reaction.

Newton's Method (for Determining the Roots of an Equation)
See **Newton–Raphson Formula**.

Newton's Rings
When a spherical lens is placed in contact with a plane plate and viewed by reflected light, circular coloured interference fringes are

seen surrounding the point of contact, the colours being most brilliant where the air film is thinnest. As the thickness of an air film between a spherical and a plane surface is easily calculated, this phenomenon gave Newton a means of determining accurately the colours produced by air films of any thickness.

Newton's Theory of Light
Newton enunciated the corpuscular theory of light in 1678. A flight of material particles was emitted by a light source and the impact of these particles on the retina produced the sensation of sight. Rectilinear propagation is a consequence of Newton's second law. Refraction and reflection were explained as due to forces of repulsion and attraction exerted, at the surface of the reflecting or refracting medium, on the particles.

Newton's 3/8th Rule
A method of computing definite integrals. It is similar to **Simpson's Rule**, except that the interval is divided into $3n$ equal parts giving

$$A = (3/8)\,h\,(y_0 + 3y_1 + 3y_2 + 2y_3 + 3y_4 + 3y_5 + 2y_6 + \dots + 3y_{3n-1} + y_{3n})$$

The notation is defined under **Simpson's Rule**.

Newton–Stirling Interpolation Formula
A modified form of the **Newton–Gauss Interpolation Formula**.

Newton, Trident of
Curve of the form

$$xy = ax^3 + bx^2 + cx + d \quad (a \neq 0)$$

Nicomedes, Conchoid of
A curve whose equation is

$$(x^2 + y^2)(x - a)^2 = x^2 b^2$$

Nordheim's Rule (for Resistivity)
The residual resistivity of a binary alloy containing a mole fraction x of element A and $(1 - x)$ of element B varies thus:

$$\rho(x) \propto x(1 - x).$$

There are significant deviations from the rule, such as when transition metals are alloyed with noble metals.

Nordheim's Rules (for Nuclear Shell Model)
Rules for the shell model of the nucleus applicable to nuclei which have odd numbers of protons and neutrons. The Nordheim number is defined as

$$N = j_{\mathrm{p}} - \ell_{\mathrm{p}} + j_{\mathrm{n}} - \ell_{\mathrm{n}}$$

where ℓ is the orbital angular momentum and j is the total angular momentum. Subscripts p and n refer to the odd proton and neutron respectively. Then the rules state:

$$\text{if} \quad N = 0, \quad I = |j_{\mathrm{n}} - j_{\mathrm{p}}|$$

$$\text{if} \quad N = \pm 1, \quad I = |j_{\mathrm{n}} - j_{\mathrm{p}}| \quad \text{or} \quad |j_{\mathrm{n}} + j_{\mathrm{p}}|$$

where I is the total angular momentum of the nucleus. There are exceptions to the rules, especially among the light nuclei.

Norris–Eyring Reverberation Formula *See* **Millington Reverberation Formula**

Norton's Theorem
A source of electrical energy can be considered as a constant-current generator in parallel with an impedance. The current can be considered as the short-circuit current and the impedance is the same as that in **Thévenin's Theorem**.

Nusselt Number
A dimensionless parameter (Nu) used for heat transfer between a solid body and a fluid flowing relative to a body.

$$(Nu) = \frac{hl}{\lambda}$$

where h is the heat-transfer coefficient (rate of loss of heat per unit area of surface and per unit temperature difference between surface and fluid), l is a characteristic linear dimension of the body and λ the thermal conductivity of the fluid. The **Biot Number** (Bi) is similarly defined but involves the thermal conductivity of the solid. The Nusselt number compares the convective capacity and the conducting capacity of the fluid, whereas the Biot number is an index of the relative resistances to heat flow in the solid and fluid.

Nyquist Rule
If a response-vector locus be plotted for a servo system for positive and negative frequencies and if the point $(-1, \mathrm{j}0)$ is not enclosed, the system is unstable.

Nyquist's Theorem
The mean-square-voltage across a resistance R in thermal equilibrium at absolute temperature T is given by

$$\overline{V}^2 = 4\,R\,k_{\mathrm{B}}T\,\Delta f$$

where Δf is the frequency band within which the voltage fluctuations are measured.

O

Occam's Razor
A maxim attributed to William (of) Occam (*ca* 1290–1350): it is vain to do with more what can be done with fewer, i.e., if the facts resulting from an experiment can be explained without assuming additional hypotheses, there are no grounds for assuming them.

Oersted *See* **Appendix**

Ohm; **Ohm**, **Acoustical**; **Ohm**, **Mechanical**; **Ohm**, **Thermal** *See* **Appendix**

Ohm's Law
In any circuit consisting of homogeneous conductors through which electricity is flowing, the ratio of the electromotive force applied across the circuit to the current flowing in the circuit is a constant if the circuit is kept under constant physical conditions.

Ohm's Law (Acoustic)
Every motion of air which corresponds to a composite set of musical tones is capable of being analysed into a sum of simple pendular (harmonic) vibrations; and to each single simple vibration corresponds a simple tone, sensible to the ear and having a pitch determined by the periodic time of the corresponding motion of the air.

Olbers' Paradox
If it were to be assumed that the universe is a static system of infinite age, Olbers deduced that every star would be absorbing as much radiation as it is emitting and that the bulk of the radiation received by a star or other body must come from distant parts of the universe. Hence the night sky would not be dark. The paradox indicates that the universe is young or is expanding.

Onsager Conductivity Equation (Debye–Hückel–Onsager Equation)

An ion moving in an electrolyte under the influence of an applied field is retarded by two effects. The relaxation or asymmetry effect is due to the ionic atmosphere of oppositely charged ions having a definite time of relaxation. The ions will also be subjected to the effect of the ionic atmosphere moving in the opposite direction to the central ion. Using these two concepts and making the assumption that the electrolyte is completely dissociated, Debye and Hückel were able to deduce an equation for the variation of equivalent conductivity of an electrolyte in any solvent in terms of the concentration. This equation was modified by Onsager to take the form

$$\Lambda = \Lambda_0 - \left\{ \frac{29{\cdot}15\,(Z_+ + Z_-)}{(\varepsilon_r T)^{1/2}\eta} + \frac{9{\cdot}90 \times 10^5}{(\varepsilon_r T)^{3/2}} \Lambda_0 \omega \right\} \sqrt{\{C(Z_+ + Z_-)\}}$$

where Z_+ and Z_- are the valencies of the ions, C is the concentration of the electrolyte in gram-equivalents/litre, Λ is the equivalent conductivity at this concentration and Λ_0 is the equivalent conductivity at infinite dilution, ε_r is the dielectric constant of the solvent and η is its viscosity.

$$\omega = Z_+ Z_- \frac{2q}{1 + \sqrt{q}}$$

where

$$q = \frac{Z_+ Z_-}{Z_+ + Z_-} \cdot \frac{\lambda_+ + \lambda_-}{Z_+ \lambda_- + Z_- \lambda_+}$$

and where λ_+ and λ_- are the mobilities (equivalent conductivities) of the ions at infinite dilution.

Onsager Equation for Dielectric Constant

Onsager has developed an approximate dielectric theory for polar substances. If the induced polarization is neglected, the dielectric constant (relative permittivity) is given by

$$\varepsilon_r = \tfrac{1}{4}\{1 + 3x + 3(1 + \tfrac{2}{3}x + x^2)^{1/2}\}$$

$$x = \frac{n\mu^2}{3k_B T} \text{ (SI)}$$

where μ is the dipole moment and n is the number of molecules or ions per unit volume.

Onsager's Reciprocal Relations
This theory, which is used in the thermodynamics of irreversible processes, states that: if a proper choice of the fluxes J_i and forces X_i has been made, the matrix of phenomenological coefficients L_{ik} is symmetrical, i.e.,

$$L_{ik} = L_{ki} \ (i, k = 1, 2, 3, \ldots n)$$

In irreversible phenomena there can be a number of causes: i.e. a concentration gradient can produce a diffusion flow; a temperature gradient a thermal flow. These causes are called forces or affinities (X) and produce fluxes or flows (J). For situations not too far removed from equilibrium, the irreversible phenomena can be expressed by linear relations of the general type

$$J_i = \sum_{k=1}^{n} L_{lk} X_k \ (l = 1, 2, 3, \ldots n)$$

Linear relations of this type are called phenomenological relations and the coefficients L_{ik} are called phenomenological coefficients.

Oppenauer Oxidation
When a secondary alcohol is refluxed with aluminium *t*-butoxide in excess acetone, it is oxidized to a ketone. This reaction is specific for secondary alcohols.

Oppenheimer–Philips Reaction
A nuclear 'stripping' process which can occur when a deuteron passes close to a nucleus. The coulombic force between the nucleus and the deuteron breaks the neutron–proton bond within the deuteron. The neutron is captured by the nucleus and the proton is repelled.

Oppenheimer–Volkoff Limit
This is the critical mass below which a star cannot become a neutron star, because the forces due to degeneracy pressure of the neutrons are greater than those due to the gravitational field. The critical mass is thought to be approximately three solar masses.

Ostwald Dilution Law

According to W. Ostwald (1888) the dissociation constant of a univalent electrolyte is given by the relation

$$k = \frac{\alpha^2 c}{1-\alpha}.$$

This law is true if activities are taken into account: i.e. the relation should be written

$$k = \frac{\alpha^2 c}{1-\alpha}\cdot\frac{f_+ + f_-}{f_\pm}$$

where c is the concentration, α is the degree of dissociation, and f_+ and f_- are the activity coefficients of the anions and cations and $f_\pm$ of the undissociated electrolyte.

Ostwald's Basicity Rule

W. Ostwald (1887) proposed an empirical relationship between the equivalent conductivity of the sodium or potassium salt of an acid and its basicity

$$\Lambda_{1024} - \Lambda_{32} = 10{\cdot}8b$$

where Λ_{1024} is the equivalent conductivity of a solution of dilution 1024 and Λ_{32} of dilution 32.

Otto Cycle

The working cycle of a four-stroke internal combustion engine consisting of inlet, compression, explosion at constant volume and expansion, and exhaust. The whole cycle corresponds to two complete revolutions of the crankshaft.

Overhauser Effect

When placed in an external magnetic field H, a substance which has nuclei of spin $\frac{1}{2}$ and magnetic moment μ_n and also unpaired electrons of spin $\frac{1}{2}$ and magnetic moment μ_e ($\mu_e < 0$) shows this effect. If an r.f. field is applied at the electron spin resonance frequency, then the ratio of the number of nuclei n_+ and n_- with spins 'up' and 'down' in the magnetic field is given by

$$n_+/n_- = \exp\{2(\mu_n - \mu_e)H/k_B T\} \approx \exp\{-2\mu_e H/k_B T\}$$

Thus, the polarization of the nuclei is as large as if the nuclei possessed the much larger electron magnetic moment. The effect arises because the nuclei interact predominantly with the electron spins, which in their turn interact thermally with the lattice.

P

Paal–Knorr Synthesis

Pyrrole derivatives may be prepared by treating a 1:4 diketone with ammonia, a primary amine or a hydrazine.

Paal–Knorr Synthesis of Thiophenes

Homologues of thiophen can be made by reaction of 1:4 diketones with phosphorus pentasulphide.

Paneth's Adsorption Rule *See* **Fajans' Precipitation Rule**

Paperitz's Equation

The most general ordinary differential equation of second order with three regular singular points. Let these points be at $z = a, b$ and c while the corresponding exponents are α, β, γ and α', β', γ' respectively. The equation is

$$\frac{d^2y}{dz^2} - \left(\frac{\alpha+\alpha'-1}{z-a} + \frac{\beta+\beta'-1}{z-b} + \frac{\gamma+\gamma'-1}{z-c}\right)\frac{dy}{dz}$$
$$+ \left\{\frac{\alpha\alpha'(a-b)(a-c)}{(z-a)^2(z-b)(z-c)} + \frac{\beta\beta'(b-a)(b-c)}{(z-a)(z-b)^2(z-c)}\right.$$
$$\left. + \frac{\gamma\gamma'(c-a)(c-b)}{(z-a)(z-b)(z-c)^2}\right\} = 0$$

where $\alpha + \alpha' + \beta + \beta' + \gamma + \gamma' = 1$. (*See* **Riemann's Symbol**.)

Pappus' Theorems

I. If an arc of a plane curve revolve about an axis in its plane, not intersecting it, the surface generated is equal to the length of the arc multiplied by the length of the path of its mean centre.

II. If a plane area revolve about an axis in its plane, not intersecting it, the volume generated is equal to the area multiplied by the length of the path of its mean centre.

Parseval's Identity
If **i**, **j**, **k**, . . . are unit orthonormal vectors in a vector space of finite dimensions, then, for any vector **u** given by $\mathbf{u} = a\mathbf{i} + b\mathbf{j} + c\mathbf{k} + \ldots$,

$$\|\mathbf{u}\|^2 = |a|^2 + |b|^2 + |c|^2 + \ldots$$

where $\|\mathbf{u}\|$ is the magnitude of **u** [given by $+(\sqrt{\mathbf{u}.\mathbf{u}})$].

Parseval's Theorem
This states that, if, in particular, $F(k)$ and $f(x)$ are **Fourier Transforms** of each other, then

$$\int_{-\infty}^{\infty} |F(k)|^2 \, dk = \int_{-\infty}^{\infty} |f(x)|^2 \, dx$$

Alternatively, the theorem is quoted with the right-hand side of the equation expressed as a sum of a series. See also **Fourier's Series**.

Pascal *See* **Appendix**

Pascal's Law
Pressure exerted at any point upon a confined liquid is transmitted undiminished in all directions.

Pascal's Limaçon (or **Pascal's Snail**)
A curve whose form is given by the parametric equations for x and y:

$$x = a\cos^2\phi + \ell\cos\phi$$
$$y = a\cos\phi\sin\phi + \ell\sin\phi$$

Pascal's Theorem
For any hexagon inscribed in a conic, the intersections of opposite pairs of sides are collinear (or at infinity).

Pascal's Triangle
A method of obtaining the coefficients of the expansion $(q+p)^n$ (binomial coefficients) from a table:

n	coefficients
1	1 1
2	1 2 1
3	1 3 3 1
4	1 4 6 4 1

Each term is derived from the sum of the two terms lying on either side of it in the line above.

Paschen–Back Effect
Paschen and Back discovered in 1912 that all the simplicity of many Zeeman patterns was lost when the magnetic separation was comparable to the natural separation of a doublet, triplet, etc. The lines appeared to influence each other until, when the magnetic separation was large compared to the natural separation, the normal **Zeeman Effect** returned both in pattern and in magnitude.

A similar effect, in which the hyperfine structure in a spectrum breaks up in a magnetic field but begins to overlap again as the field is further increased, is called the **Back–Goudsmit Effect**.

Paschen Series *See* **Balmer Series**

Paschen's Law
At a constant temperature, the breakdown voltage of a gas is a function only of the product of the gas pressure and the distance between parallel plane electrodes.

Patterson Function
In the determination of crystal structures by x-ray analysis, a lack of knowledge of the phases of structure factors makes it impossible directly to compute an electron density map and hence to show the positions of the atoms within a unit cell. Patterson suggested the use of the function

$$P(\mathbf{r}) = \frac{1}{V}\sum_{h} |F_{\mathbf{h}}|^2 \exp(-2\pi j\mathbf{h}.\mathbf{r})$$

where $F_{\mathbf{h}}$ are structure factors, $\mathbf{h}$ is a reciprocal-lattice-point coordinate, $\mathbf{r}$ is a real-lattice coordinate and V is the volume of the unit cell. Using **Friedel's Law** and **Fourier Transforms**, it can be shown to be equivalent to the alternative form

$$P(\mathbf{r}) = V\int_{V} \rho(\mathbf{u})\rho(\mathbf{u}+\mathbf{r})\,\mathrm{d}v$$

where ρ is the electron distribution function, and integration is over the complete unit cell. $\mathbf{u}$ and $\mathbf{u}+\mathbf{r}$ represent displaced coordinates within the unit cell. Electron density maps can be obtained from the function in this form.

Pauli Exclusion Principle
If individual systems belong to a certain class, e.g. electrons, neutrons, protons, only antisymmetric functions may be used to describe the assemblage. This principle, in particular, states that it is impossible for any two electrons in the same atom to have four identical quantum numbers.

Pauli Spin Matrices
These are denoted

$$\sigma_x = \begin{pmatrix} 0 & 1 \\ 1 & 0 \end{pmatrix} \quad \sigma_y = \begin{pmatrix} 0 & -\mathrm{j} \\ \mathrm{j} & 0 \end{pmatrix} \quad \sigma_z = \begin{pmatrix} 1 & 0 \\ 0 & -1 \end{pmatrix}$$

$$\sigma^2 = \begin{pmatrix} 3 & 0 \\ 0 & 3 \end{pmatrix} \quad \text{where} \quad {\sigma_x}^2 + {\sigma_y}^2 + {\sigma_z}^2 = \sigma^2.$$

The total spin (intrinsic angular momentum) of an electron is then expressed as $S = \frac{1}{2}\hbar\sigma$.

Pauli Spin Susceptibility
The paramagnetic susceptibility of a free electron gas. It arises from the ability of electrons within an energy of approximately $k_B T$ of the **Fermi Energy** to alter their direction of spin in a magnetic field.

Peano's Axioms
These axioms define the properties of positive integers:

(1) 1 is a positive integer.
(2) Each positive integer n has a unique successor denoted $S(n)$.
(3) $S(n) \neq 1$.
(4) If $S(n) = S(m)$ then $n = m$.
(5) The principle of finite induction holds: i.e. the set of positive integers 1 plus the successors $S(n)$ for each n contains all positive integers.

Pearl–Reed Curve
A curve of the form

$$y = k/(1 + \mathrm{e}^{a+bx}) \quad \text{where} \quad b < 0.$$

Pearson's Distributions
Frequency functions $\phi(x)$ of many continuous distribution functions, for instance **Gauss' Error Curve**, can be expressed in the form

$$\frac{1}{\phi}\frac{\mathrm{d}\phi}{\mathrm{d}x} = \frac{x-a}{b_0+b_1x+b_2x^2}$$

The parameters a, b_0, b_1 and b_2 define each distribution, and the distributions are classified according to the roots of the equation

$$b_0+b_1x+b_2x^2 = 0.$$

Péclet's Number
A dimensionless number for a fluid flowing past a body and equal to **Prandtl's Number** × **Reynolds Number**.

Peierls Stress
The minimum energy required to remove a dislocation from its rest position of minimum energy.

Pellian Equation
An equation of the form

$$x^2 - Ay^2 = 1$$

where A is a positive integer that is not a perfect square.

Peltier Effect
Heat is, in general, absorbed or liberated when an electric current crosses the junction between two metals A and B. The heat is proportional to the current i and is equal to Πi, where Π is called the Peltier coefficient of A with respect to B and varies with temperature.

Penning Effect
The ionization of one species of gas molecule by collision with another species in a mixture of gases, when the former species is in a metastable excitation level which is above the ionization potential of the other.

Penrose Process
A method of extracting energy from a rotating black hole. A particle must spiral just into a black hole, in a direction opposite to the black

hole's rotational direction, and break into two fragments. One of these fragments then leaves the black hole with greater energy than the ingoing particle.

Penrose's Theorem

A collapsing object whose radius is less than its gravitational radius must collapse into a singularity. The gravitational radius of an object is the radius which that object would have, such that light emitted from its surface just ceases to escape because of the gravitational field at that distance.

Perkin Reaction

When an aromatic aldehyde is heated in the presence of its sodium salt, with an anhydride of an aliphatic salt containing two α-hydrogen atoms, condensation takes place to form a β-aryl acrylic acid.

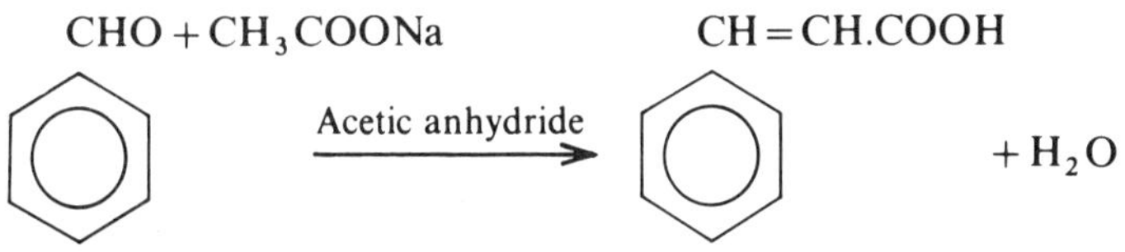

Petzval Surface

The paraboloidal surface over which an image is formed by a lens system in the absence of astigmatism.

Pfaffian Differential Equation

A first-order differential equation of the form

$$A(x, y, z)\mathrm{d}x + B(x, y, z)\mathrm{d}y + C(x, y, z)\mathrm{d}z = 0.$$

It is integrable if, and only if, there is an integrating factor $\mu = (x, y, z)$ such that $\mu(\mathrm{A}\mathrm{d}x + \mathrm{B}\mathrm{d}y + \mathrm{C}\mathrm{d}z)$ is an exact differential.

Pfitzinger Reaction

Derivatives of cinchonic acid (4-quinoline carboxylic acid) may be prepared by the reaction of isatin with aldehydes and ketones.

Pfund Series *See* **Balmer Series**

Picard's Method

A method of solving ordinary differential equations by successive approximations. If

$$dy/dx = f(x, y)$$

a solution may be found of the form

$$y = y_0 + \int_{x_0}^{x} f(x, y)dx \equiv y_0 + \int_{x_0}^{x} \frac{dy}{dx}dx$$

where $y = y_0$ at $x = x_0$. If the approximation $y = y_0$ under the integral sign is made, the integral is suitable for numerical integration as it is a function of x and y_0 only. Thus,

$$y_1 = y_0 + \int_{x_0}^{x} f(x, y_0)\, dx$$

The process is repeated to give

$$y_2 = y_0 + \int_{x_0}^{x} f(x, y_1)\, dx$$

and so on.

Pippard Coherence Length

The range of coherence, ξ, of the wave-functions of superconducting electrons in a superconductor. Its value at zero temperature is taken as

$$\xi_0 = \frac{hv_F}{\pi^2 E_g}$$

where E_g is the energy gap obtained from the theory of **Bardeen, Cooper and Schrieffer** and v_F is the velocity of electrons at the **Fermi Surface**. In tin, ξ is approximately 1 μm. In applying the **London Equation**, if the vector potential is varying rapidly with position, it is necessary to consider the current density at a point as proportional to an average of the vector potential over a region of characteristic size ξ.

Planck *See* **Appendix**

Planck Function
A thermodynamic function Y given by

$$Y = -(U/T) - (pV/T) + S$$

where U is internal energy and S entropy. It is equal to minus the quotient of the **Gibbs' Function** and the temperature T.

Planck's Constant *See* **Appendix**

Planck's Law
Arising from **Planck's Quantum Theory**, Einstein postulated that the energy content E of an energy packet or photon can be related to its frequency ν by

$$E = h\nu.$$

This is often referred to as Planck's Law.

Planck's Quantum Theory
The Rayleigh–Jeans formula for the distribution of energy in the spectrum of black-body radiation is:

$$E = \frac{8\pi k_B T}{\lambda^4}$$

indicating a steady increase of energy density E with decreasing wavelength λ as for the dotted curve in the figure. The experimental curve is shown by the full line and has a maximum for a given value of

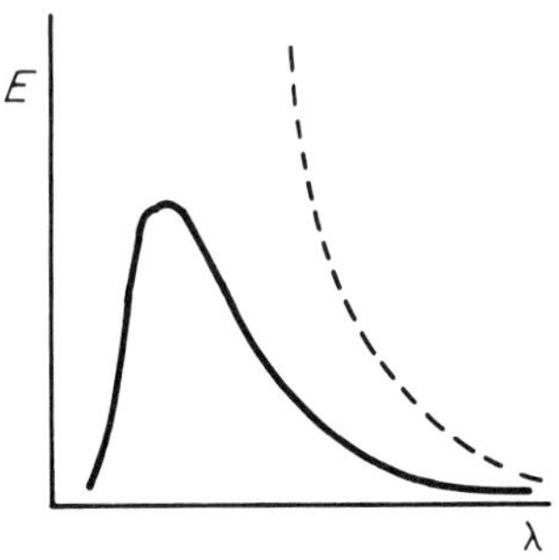

Planck's Quantum Theory

λ depending upon the temperature of the black body. This discrepancy was resolved by Planck, who showed that the error lay in applying the classical laws of mechanics to atomic particles. These observations were the basis of the modern quantum theory. In classical theory the amplitude of vibration of a linear oscillator may have any desired value within certain limits. Planck assumed, however, that the amplitude of vibration could only assume certain discrete values,

$$E = 0, h\nu, 2h\nu, \ldots$$

Planck's Radiation Formula *See* **Planck's Quantum Theory**

Plateau Border
Also called the **Gibbs Ring**, this is a border region in a soap film in contact with a wire frame. In this region there is a lower pressure than the surrounding atmospheric pressure and the curvature is concave outwards. This type of region also occurs at the intersection of three soap films.

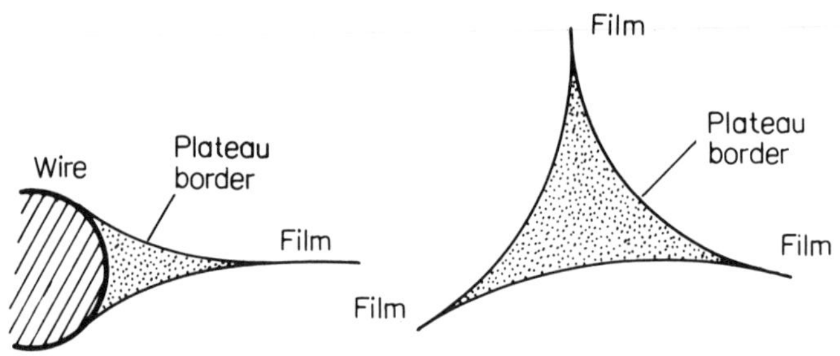

Plateau Problem
The determination of the minimum area contained by a boundary.

Platonic Polyhedra
These are the five possible regular solids, the tetrahedron (four faces), cube (six faces), octahedron (eight faces), dodecahedron (twelve faces) and icosahedron (twenty faces).

Playfair's Axiom
Two intersecting straight lines cannot be parallel to a third straight line.

Pockels Effect
If a light beam is passed through certain non-centrosymmetric crystals in the same direction as an applied electric field, the refractive indices of the crystals change with change of applied field, and the crystals become birefringent, the difference in velocity for the two rays being proportional to the electric field. If a plane polarized beam of light is used, the plane of polarization is rotated as the light passes through the crystal. The effect is analogous to the **Faraday Effect** in a magnetic field. The effect can be used in fibre-optical communications to modulate an incident laser beam. *See also* **Kerr Electro-optic Effect**.

Poise *See* **Appendix**

Poiseuille's Formula
The volume of liquid V, flowing per second through a capillary tube of length l and radius r under a pressure p is given by

$$V = \frac{\pi p r^4}{8 l \eta}$$

where η is the viscosity of the liquid.

Poisson Distribution Law
The frequency distribution for the number of random events r in a constant time interval is given by

$$\mathrm{P}(r) = \frac{m^r \mathrm{e}^{-m}}{r!}$$

where m is the mean number of counts in the time interval.

Poisson's Bracket
A differential operator on two (twice continuously differentiable) functions U and V describing a physical state:

$$[\mathrm{U}, \mathrm{V}] = \sum_{m=1}^{n} \left[\frac{\partial \mathrm{U}}{\partial p_m} \frac{\partial \mathrm{V}}{\partial q_m} - \frac{\partial \mathrm{U}}{\partial q_m} \frac{\partial \mathrm{V}}{\partial p_m} \right]$$

where p and q are conjugate variables, e.g. momenta and coordinates (cf. **Hamiltonian**) and n is the number of degrees of freedom of the system.

The **Heisenberg Uncertainty Principle** can be stated thus: If the Poisson bracket $[U, V] = K$, where K is a constant independent of the coordinate system used, the product of the uncertainties in simultaneous measurement of U and V, i.e. ΔU and ΔV respectively, is given by

$$\Delta U \Delta V = Kh$$

where h is Planck's constant.

If W is a similar function depending on p and q, then

$$\Big[U, [V, W]\Big] + \Big[V, [W, U]\Big] + \Big[W, [U, V]\Big] = 0$$

and this is called **Poisson's identity**.

Poisson's Equation

If, within a closed surface, there exists a volume density of charge ρ the potential V on the surface will be given in SI units by

$$\nabla^2 V = \frac{\partial^2 V}{\partial x^2} + \frac{\partial^2 V}{\partial y^2} + \frac{\partial^2 V}{\partial z^2} = \frac{\rho}{\varepsilon}$$

where ε is the permittivity within the closed surface. This determines the distribution of charge within a surface when the potential at every point on the surface is known.

Poisson's Integral Formula (for Bessel Functions)

If $J_n(x)$ is a **Bessel Function**, then

$$J_n = \frac{2(x/2)^n}{\sqrt{\pi}\,\Gamma(n+\frac{1}{2})} \int_0^{\pi/2} \cos(x \cos t) \sin^{2n} t \, dt$$

$$(n > -\tfrac{1}{2})$$

Poisson's Law

When a gas expands adiabatically

$$pV^{\gamma} = \text{constant}$$

where p is the pressure, V is the volume and $\gamma = C_p/C_V$ the ratio of the specific heats of a gas.

Poisson's Ratio
The ratio of lateral strain to longitudinal strain when a body is in simple tension or compression.

Poisson's Sum Formula
A method of evaluating a sum of the form

$$\sum_{n=-\infty}^{\infty} \mathrm{f}(an)$$

Let $\mathrm{F}(k)$ be the **Fourier Transform** of $\mathrm{f}(x)$. The sum is equal to

$$\frac{\sqrt{(2\pi)}}{a} \sum_{n=-\infty}^{\infty} \mathrm{F}\left(\frac{2n\pi}{a}\right)$$

which may be easier to evaluate.

Poisson Trials
In probability theory these are a series of independent trials: i.e. 'success' on any trial is independent of all preceding trials.

Polonovski Reaction
The removal of a methyl group can be accomplished by this reaction. A tertiary amine is converted to the amine oxide, which reacts with acetic anhydride to give formaldehyde and an amide which can be hydrolysed to the secondary amine.

Poncelet *See* **Appendix**

Poncelet's Formula
A method of computing integrals similar to **Simpson's Rule** but less accurate.

$$A = \frac{h}{2}\left\{\frac{y_0 + y_n}{4} - \frac{y_1 + y_{n-1}}{4} + 2y_1 + 2y_3 + \ldots + 2y_{n-1}\right\}$$

where the notation is that used in Simpson's rule.

Poole–Frenkel Effect
If a semiconductor, either crystalline or amorphous, contains donors, then a field E lowers the ionization energy of the centres by $2(e^3/\varepsilon_r)^{1/2}$ where ε_r is the background dielectric constant. The formula is obtained from a one-dimensional analysis and assumes no tunnelling.

Portevin–Le Chatelier Effect
The continually repeating non-smooth deformation of a specimen when subjected to a uniformly increasing stress. In polycrystalline material the initial steps are due to yielding inside the single-crystal grains, but the later steps are due to the presence of grain boundaries.

Pourbaix Diagram
In any electrode reaction, electrode potentials can be plotted against pH. If this is done for various activities of the reactants, a diagram can be constructed in which different areas or domains can be associated with different behaviours of the reactants. Thus, for the behaviour of iron in aqueous solution, a diagram can be constructed as shown where the domains correspond to

A. Passivation if the oxide stable in this area is protective.
B. Corrosion. Iron ions are stable in this area.
C. Immunity due to iron becoming cathodic.
D. Corrosion due to formation of iron anions.

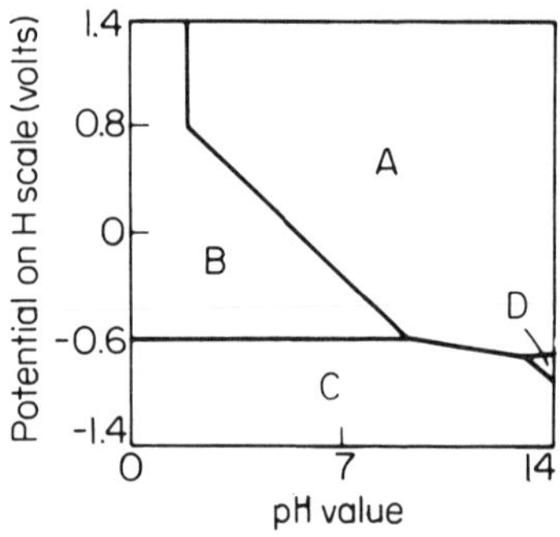

Poynting–Robertson Effect
The spiralling inwards of a small orbiting particle in the vicinity of the sun or similar rotating body. The effect arises because radiation falls preferentially on the leading edge and produces drag.

Poynting's Theorem
When energy is transmitted through an electromagnetic field, if a small volume is enclosed by a surface in the field, and if the energy of electric strain and magnetic flux contained in it be varying, the amount of energy which enters each element of the surface is measured by the sum of the products of the electric and magnetic forces resolved along each element of the surface, multiplied by the sine of the angle between their directions using SI units.

The vector measuring the rate of flow of electromagnetic energy per unit area is known as **Poynting's vector**.

Prandtl's Number
A dimensionless number (Pr) for a fluid defined as

$$(Pr) = \frac{C_p \eta}{\lambda \rho}$$

where η is the dynamic viscosity of the fluid, C_p is its specific heat at constant pressure, λ its thermal conductivity and ρ the density.

Preece *See* **Appendix**

Preston's Rule
In the anomalous **Zeeman Effect**, lines of the same series show the same pattern.

Prévost Theory of Heat Exchange
In the eighteenth century, ideas regarding radiant energy were confused. Radiations were described as hot and cold, e.g. a block of ice emitted cold radiation. Prévost recognized the looseness of this method of description in 1792, and said that all bodies emit radiant energy, the amount of which increases with temperature and is not affected by surrounding bodies. The rise or fall of temperature of a body is due to the exchange of radiant energy with other surrounding bodies.

Prileschaiev Reaction
Peracids convert olefins into olefin oxides.

$$RCH = CHR + C_6H_5COO_2Na \rightarrow R.\overset{\quad O \quad}{CH - CH}.R + C_6H_5COONa$$

Usually perbenzoic or monoperphthalic acids are used in this reaction.

Primakoff Effect
The production of a neutral meson as a result of the interaction of an incident gamma ray with a virtual gamma ray of the Coulomb field of the nucleus.

Prince Rupert's Cubes
The largest cube which can be made to pass through a channel within a given cube of side a is one of side $\frac{3}{4}\sqrt{2}a$ (i.e. 1.06065 a).

Pringsheim's Theorem (for a Convergent Series)
If Σu_n converges and $u_n \to 0$ monotonically, then $nu_n \to 0$.

Prins Reaction
Arylethylenes react with paraformaldehyde, in acetic acid solution in the presence of concentrated sulphuric acid, to give unsaturated alcohols, glycols or their acetates.

Proca Equation
A relativistic equation for a form of vector potential describing a hypothetical particle of unit spin. *See* **Dirac Equation** *and* **Klein–Gordon Equation**.

Procopiu Effect
The appearance, in a coil wound coaxially around a ferromagnetic wire, of an electromotive force which has double the frequency of an alternating current passed through the wire.

Proust's Law
All compounds contain elements in certain definite proportions and no others, regardless of conditions of production. It is also called the law of definite proportions.

Prout's Hypothesis
W. Prout in 1815 suggested that all atoms were built up from the same unit, the hydrogen atom. The fact that not all atomic weights were integral led to this hypothesis being regarded as unsound. The discovery of isotopes removed this objection, and a modified form of the hypothesis, utilizing the proton or hydrogen nucleus and the neutron as the building units of all atomic nuclei, has gained universal acceptance.

Prym's Functions

$$P(x;\rho) = \int_0^{\rho} t^{x-1} \mathrm{e}^{-t}\,\mathrm{d}t$$

and

$$Q(x;\rho) = \int_{\rho}^{\infty} t^{x-1}\mathrm{e}^{-t}\,\mathrm{d}t$$

are known as Incomplete Gamma Functions. The special functions which arise when $\rho = 1$ are called Prym's Functions.

Pschorr Synthesis
A method of synthesizing phenanthrene. *o*-nitrobenzaldehyde is condensed with sodium phenylacetate by means of the **Perkin Reaction**. The product α-phenyl *o*-nitrocinnamic acid is reduced and diazotized. The azo-compound is treated with sulphuric acid and copper powder to give phenanthrene-9-carboxylic acid. This compound loses carbon dioxide when heated strongly, and phenanthrene is formed. The synthesis offers means of preparing substituted phenanthrenes.

Ptolemy's Theorem
The rectangle contained by the diagonals of a quadrilateral inscribed in a circle is equal to the sum of the two rectangles contained by its opposite sides.

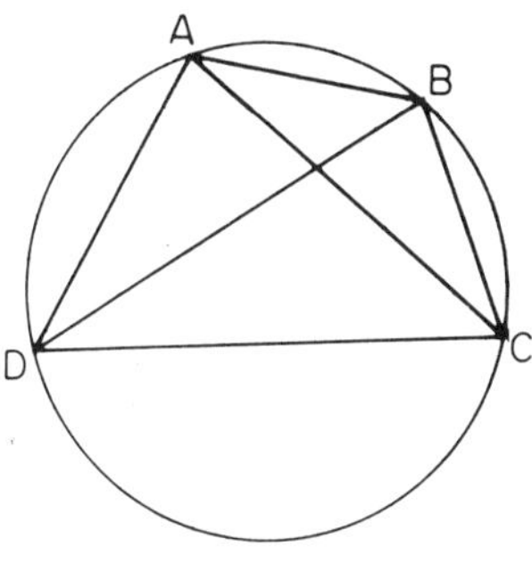

AC·BD = AD·BC + AB·CD

Ptolemy's Theorem

Purdie Method of Alkylation
When an alkyl halide is heated with an alcohol in the presence of silver oxide, an ether is formed with the elimination of water and silver iodide.

Purkinje Effect
The human eye is more sensitive to blue light than to yellow light when illumination is poor (less than 1 lumen m^{-2}) and to yellow rather than blue when the illumination is good.

Pythagoras' Theorem
In a right-angled triangle the square on the hypotenuse is equal to the sum of the squares on the other two sides.

Pythagorean Identity
The trigonometric identity

$$\sin^2 x + \cos^2 x = 1.$$

Pythagorean Numbers
Any set of integers satisfying the equation $x^2 + y^2 = z^2$. They are given by $a^2 - b^2$, $2ab$ and $a^2 + b^2$, where a and b are any integers.

Q

Quincke Effect
A substance of high magnetic susceptibility tends to move into the strong region of a magnetic field. Similarly, a substance of high dielectric constant tends to move into a strong electric field.

R

Raabe's Test for Convergence
The positive term series Σu_n is convergent if

$$n\left(\frac{u_n}{u_{n+1}} - 1\right) \geqslant \rho \qquad \text{where } \rho > 1.$$

It is divergent if

$$n\left(\frac{u_n}{u_{n+1}} - 1\right) \leqslant 1.$$

Raman Scattering
If any substance, gaseous, liquid or solid is exposed to radiation of a definite frequency, the light scattered at right angles to the incident beam contains a spectral distribution of frequencies characteristic of the substance under investigation. Such a spectrum is known as the **Raman spectrum**. It arises because of interaction with molecular vibrations in the substance. A most important phenomenon recently observed is coherent Raman scattering. It has been observed that if certain substances are excited by coherent radiation from a laser, the Raman spectrum contains coherent components.

Ramsauer Effect
For low-energy electrons passing through a gas, there is a peak in the absorption cross-section at a particular electron energy. The peak corresponds to an energy at which electrons are particularly efficient at ionizing or exciting atoms in the gas. Ramsauer first discovered the effect in the gases xenon, krypton and argon.

Ramsay–Young Rule
If two substances have the same vapour pressure p at temperatures T_A

and T_B and p' at T_A' and T_B' respectively, Ramsay and Young in 1885 showed empirically that

$$\frac{T_A}{T_B} - \frac{T_A'}{T_B'} = C(T_A - T_A')$$

where C is a constant.

Rankine Cycle
A steam engine cycle, characterized by the introduction of water at boiler pressure by a pump; evaporation; adiabatic expansion to condenser pressure; and condensation to the initial point.

Rankine Efficiency
The efficiency of an ideal engine working under a **Rankine Cycle**.

Rankine–Hugoniot Relations
Equations expressing the conservation of mass, momentum and energy on either side of the discontinuity in a shock wave.

$$\rho_1 v_1 = \rho_2 v_2$$

$$p_1 + \rho_1 {v_1}^2 = p_2 + \rho_2 {v_2}^2$$

$$\tfrac{1}{2}{v_1}^2 + U_1 + \frac{p_1}{\rho_1} = \tfrac{1}{2}{v_2}^2 + U_2 + \frac{p_2}{\rho_2}$$

Subscripts 1 and 2 refer to fluid immediately to the right and left of the discontinuity respectively. p is pressure, ρ density, v flow velocity and U internal energy. The **Hugoniot function** $\mathscr{H}$ is defined by

$$\mathscr{H} = U_2 - U_1 - \tfrac{1}{2}\left(p_2 + p_1\right)\left(\frac{1}{\rho_1} - \frac{1}{\rho_2}\right)$$

and, by elimination of the flow velocities from the Rankine–Hugoniot relations, it can be shown that where these equations are applicable the Hugoniot function is zero.

Rankine's Formula
An empirical formula for the collapsing load for a column

$$L = \frac{\sigma A}{1 + a(l/k)^2}$$

where L is the collapsing load, A is the cross-sectional area of the column, σ is the safe compressive stress, l the length of column and k the least radius of gyration of section. a is given by $\sigma/\pi^2 E$, where E is **Young's Modulus**.

Rankine Temperature Scale
A scale of temperature with the freezing point of water as 491·7° and the boiling point of water as 671·7° at normal pressure.

Raoult's Law
If p_0 is the vapour pressure of a pure solvent and p the vapour pressure of a substance containing n_2 moles of solute in n_1 moles of solvent, F. Raoult showed that

$$\frac{p_0 - p}{p} = \frac{n_2}{n_1 + n_2} = x$$

where x is the mole fraction of solute.

Raschig Process
Phenyl chloride is produced commercially by passing benzene vapour, air and hydrogen chloride over cuprous chloride as catalyst.

Ray; Rayl; Rayleigh *See* **Appendix**

Rayleigh Criterion
Two patterns (for instance **Fraunhofer Diffraction** patterns or **Airy's Discs**) of equal intensity may be said to be resolved when the central maximum of one pattern falls over the minimum of the other.

Rayleigh–Jeans Radiation Formula *See* **Planck's Quantum Theory**

Rayleigh–Ritz Method
A variational method of obtaining approximate solutions of differential equations by assuming arbitrary parameters in trial functions and continuously improving them by iteration.

Rayleigh Scattering
A process of scattering of electromagnetic radiation in which there is no change of wavelength and in which the intensity of the scattered radiation (usually in the optical region) is proportional to ν^4, where ν

is the frequency of the radiation. If the scattering centres can be represented by dipole oscillators of natural frequency ν_0, then, if $\nu \ll \nu_0$, the intensity of the scattered light is given in SI units by

$$I_s = \frac{\pi N e^4 v^4}{6m^2 c^2 v_0{}^4} I_0$$

where I_0 is the intensity of the primary radiation and N is the number of oscillators of mass m and charge e. If $\nu \gg \nu_0$, **Thomson Scattering** occurs.

Alternatively, if the scattering is considered to be by N dielectric spheres of radius r where $r \ll \lambda$ (λ the wavelength of the light), then the intensity of the scattered light at an angle θ to the incident radiation and distance d from the scattering centres is given by

$$I_s = \frac{9}{2} \frac{\pi^2 N}{d^2} \left(\frac{n^2 - 1}{n^2 + 2}\right)^2 \frac{V^2}{\lambda^4} (1 + \cos^2 \theta) I_0$$

where n is the refractive index of the spheres relative to the surrounding medium and V is the volume of spheres. *See also* **Tyndall Effect.**

Rayleigh's Equation of Group Waves

$$v' = v - \lambda \frac{dv}{d\lambda}$$

This relation is known as Rayleigh's Equation, where v' is the group velocity of a wave group, v is the wave velocity and λ is the wavelength.

Rayleigh Waves

In a solid elastic medium of finite size, a disturbance will produce surface waves in addition to waves moving through the bulk material. Rayleigh waves are surface waves whose vibrations are in a plane perpendicular to the surface and containing the direction of propagation. Where the vibrations are horizontal and perpendicular to the direction of propagation, the waves are called **Love waves**. Surface waves are attenuated less with distance than waves propagated through the bulk material and are of particular importance, therefore, at large distances from the source.

Réaumur Scale
A scale of temperature with 0° at the freezing point and 80° at the boiling point of water at normal pressure.

Reformatsky Reaction
An α- or β-bromacid ester will react with carbonyl compound in the presence of zinc to form a β- or γ-hydroxy ester. The reaction is carried out by adding zinc to a mixture of the bromacid ester and the carbonyl compound.

Regge Poles
T. Regge (1959) devised a mathematical procedure for obtaining asymptotic bounds on scattering amplitudes, with particular application to the theory of elementary particles and their interaction. Regge poles are mathematical singularities which arise from the theory, and are an important feature.

Rehbinder Effect
The decrease in hardness and strength of ionic crystals with the adsorption of surface-active molecules.

Reimer–Tiemann Reaction
o-hydroxybenzaldehyde, contaminated with a little of the *p*-isomer, may be prepared by refluxing an alkaline solution of phenol with chloroform. The excess chloroform is distilled off and the residual liquid acidified. Replacement of the chloroform by carbon tetrachloride leads to a phenolic acid instead of the aldehyde.

Reyn *See* **Appendix**

Reynolds Number
When a number of geometrically similar bodies fall through a fluid of viscosity η and density ρ with a velocity v, the Reynolds number is, for any one of the bodies, $R = \rho l v/\eta$, where l is the length of a corresponding linear dimension. The specific resistance to motion for a number of geometrically similar bodies in a fluid is the same in all cases at velocities for which the Reynolds number has the same numerical value.

Riccati–Bessel Functions
The differential equation

$$x^2(\mathrm{d}^2z/\mathrm{d}x^2)+[x^2-n(n+1)]z=0$$

($n=0, \pm 1, \pm 2, \ldots$) has a general solution of the form:

$$\begin{aligned} z &= S_n(x)+C_n(x) \\ &= \sqrt{(\pi x/2)}\,\mathrm{J}_{n+1/2}(x)+\sqrt{(\pi x/2)}\,\mathrm{N}_{n+1/2}(x) \end{aligned}$$

where $S_n(x)$ and $C_n(x)$ are independent solutions called Riccati–Bessel functions. *See also* **Bessel Functions**.

Riccati's Equation
A first-order non-linear differential equation of the form

$$\mathrm{d}y/\mathrm{d}x=\mathrm{f}_1(x)y^2+\mathrm{f}_2(x)y+\mathrm{f}_3(x)$$

Richardson–Dushman Equation
In the emission from a metallic thermionic cathode

$$i=A(1-r)T^2\mathrm{e}^{-b/T}$$

where i is the saturation current density, A is a constant (equal to 120 A $\mathrm{cm}^{-2}\mathrm{K}^{-2}$ for i in A cm^{-2}), b is the absolute temperature equivalent of the work function and r is a reflection coefficient to allow for irregularities of the surface.

Richardson Effect *See* **Barnett Effect**

Richards' Rule
The molar heat of fusion of a solid, divided by its melting point, is equal to approximately 2. The rule is analogous to **Trouton's Law** for vaporization, but is less generally applicable.

Riemann–Hugoniot Catastrophe *See* **Thom's Classification Theorem**

Riemann Integral
This is a special case of the **Lebesgue Integral** but does not necessarily

exist, even if the Lebesgue integral does. The Riemann integral is defined as

$$\int_a^b f(x)\,dx \equiv \lim_{\substack{n\to\infty \\ \max \Delta_j x \to 0}} \sum_{j=1}^{n} f(\bar{x}_j)\,\Delta_j x$$

where $\{\Delta_j x\} = \{x_j - x_{j-1}\}$ is a subdivision of the interval $[a, b]$, and $x_{j-1} \leqslant \bar{x}_j \leqslant x_j$. Also $\sum_{j=1}^{n} \Delta_j x = (b-a)$ and the integral is defined, provided a limit exists for this sum for all possible subdivisions and for any choice of $\{\bar{x}_j\}$.

Riemann's Lemma

If $\phi(x)$ is non-decreasing and bounded in the range $b > x > a$, and λ is large, then

$$\int_a^b \phi(x)\cos(\lambda x)\;dx \quad \text{and} \quad \int_a^b \phi(x)\sin(\lambda x)\,dx$$

are $O(1/\lambda)$.

Riemann's Mapping Theorem

Any simply connected domain with at least two boundary points can always be mapped conformally on to a half plane or on to a circular disc.

Riemann's Surfaces

A geometrical representation of multi-valued functions in the complex plane which is developed into two or more interleaving surfaces.

Riemann's Symbol

A shorthand notation for a general solution of an ordinary differential equation with three regular singular points. For **Paperitz's Equation** this takes the form

$$y = P\begin{pmatrix} a & b & c & \\ \alpha & \beta & \gamma & z \\ \alpha' & \beta' & \gamma' & \end{pmatrix}$$

The top three terms (a, b, c) give the positions of the regular singular points in the complex z plane. The other two rows give corresponding values of the indices (i.e. solutions of the indicial equation). Thus at a, the indices are α and α'. The z in the fourth column denotes the independent variable.

Riemann Zeta Function
Defined by either

$$\zeta(z) = \sum_{n=1}^{\infty} n^{-z}$$

or

$$\zeta(z) = \frac{1}{\Gamma(z)} \int_0^{\infty} \frac{x^{z-1}}{e^x - 1}\, dx$$

where $z = x + jy$ and $x > 1$.

Righi–Leduc Effect
If a temperature gradient dT/dx is established along a conductor or semiconductor, and if a magnetic field is applied orthogonal to this temperature gradient, then a further temperature gradient dT/dy is established at right angles to both the magnetic field and to the first temperature gradient, such that

$$\frac{dT}{dy} = S\, B_z \frac{dT}{dx}$$

where S is the Righi–Leduc coefficient and B_z is the magnetic flux density.

Rinmann's Green Reaction
A reaction for zinc, based on the fact that, when cobalt oxide is ignited with zinc oxide, a green spinel-type double oxide is formed.

Ritter Reaction
Tertiary carbonium ions formed from alkenes or alcohols add to nitriles to give nitrilium salts. Addition of water to this salt gives a tertiary alkyl amide.

Ritz Combination Principle
The wave-number of any spectral line in a given series may be represented by the difference of two terms:

$$\tilde{\nu} = \frac{R}{x^2} - \frac{R}{y^2}.$$

x remains constant for any given series; y assumes different integral values to give the lines in that series. R is a constant known as the **Rydberg Constant** (*see* **Appendix**).

Robert's Law
For every mechanical linkage there are at least two substitutes that will produce the same desired motion. Furthermore, the two alternative linkages are related to the first by a series of similar triangles.

Roche Limit
The minimum distance at which a small body can rotate in a circular Keplerian orbit, about a central mass, without breaking up under the influence of gravitational tension.

Rodrigues' Formula

$$\mathrm{P}_m(x) = \frac{1}{2^m m!}\frac{\mathrm{d}^m}{\mathrm{d}x^m}(x^2 - 1)^m$$

P_m is the **Legendre Coefficient**.

Rolle's Theorem
Suppose $\mathrm{f}(x)$ is continuous in the closed interval (a, b) and has a derivative $\mathrm{f}'(x)$ for every x, such that $a < x < b$.

If $\mathrm{f}(a) = 0$ and $\mathrm{f}(b) = 0$, then $\mathrm{f}'(x) = 0$ for at least one value of x between a and b.

Röntgen *See* **Appendix**

Röntgen Rays
An alternative name for x-rays.

Rosenmund Reduction
Aldehydes may be prepared by the reduction of an acid chloride with hydrogen in boiling xylene, using a suspension of palladized barium sulphate as catalyst. Reduction of the aldehyde to the alcohol does not occur, as barium chloride inhibits this reaction.

Routh's Rule
A rule for finding the roots of an equation, whose real parts are positive, by inspection of variations of signs of matrices of its

coefficients. If $f(x) = a_0x^n + a_1x^{n-1} + \ldots a_{n-1}x + a_n = 0$ is the equation and

$$M_0 = a_0 > 0 \qquad M_1 = a_1 \qquad M_2 = \begin{vmatrix} a_1 & a_0 \\ a_3 & a_2 \end{vmatrix}$$

$$M_3 = \begin{vmatrix} a_1 & a_0 & 0 \\ a_3 & a_2 & a_1 \\ a_5 & a_4 & a_3 \end{vmatrix} \qquad M_4 = \begin{vmatrix} a_1 & a_0 & 0 & 0 \\ a_3 & a_2 & a_1 & a_0 \\ a_5 & a_4 & a_3 & a_2 \\ a_7 & a_6 & a_5 & a_4 \end{vmatrix}$$

and so on, then the number of roots equals the number of sign changes in one of sequences

$$M_0,\ M_1,\ M_2/M_1,\ M_3/M_2, \ldots M_n/M_{n-1}$$

or

$$M_0,\ M_1,\ M_1M_2,\ M_2M_3, \ldots M_{n-1}M_{n-2},\ a_n.$$

Routh's Rule of Inertia

The moment of inertia I of a body about an axis of symmetry is given by:

$$I = \frac{M(a^2 + b^2 + c^2)}{d}$$

where M is the mass and a, b and c the lengths of the perpendicular semi-axes of the body. $d = 3, 4$ or 5 according to whether the body is a rectangular parallelepiped, an elliptic cylinder or an ellipsoid.

Ruff Degradation

A method of converting aldohexoses to the corresponding aldopentose. The hexose is oxidized to the aldonic acid of which the calcium salt is then formed. This salt, on decarboxylation treatment with a mixture of hydrogen peroxide and ferrous acetate (Fenton's reagent), yields the pentose.

Ruggli Principle

According to this principle, by using sufficiently dilute solutions of hydroxyacids, the distance between the different molecules can be made greater than the distance between the hydroxyl group and carboxyl group in the same molecule. The formation of a cyclic lactone will be preferred, therefore, to the condensation of two molecules together to form a long chain.

Runge–Kutta Method (of Numerical Integration)
A widely used method of computing definite integrals in which fixed increments h of the independent variable x are considered. For a first-order equation of the form

$$y' = \mathrm{f}(x, y)$$

the following are obtained:

$k_1 = \mathrm{f}(x_0, y_0)h$ where x_0 and y_0 are initial values of x and y

$$k_2 = \mathrm{f}(x_0 + h/2, y_0 + k_1/2)h$$
$$k_3 = \mathrm{f}(x_0 + h/2, y_0 + k_2/2)h$$
$$k_4 = \mathrm{f}(x_0 + h, y_0 + k_3)h.$$

This gives values of x and y, after the first increment, of

$$x_1 = x_0 + h \quad \text{and} \quad y_1 = y_0 + \Delta y$$

where

$$\Delta y = (k_1 + 2k_2 + 2k_3 + k_4)/6.$$

Values of x and y after further increments are obtained in a similar way. The error is O (h^5). The method can also be applied to second-order equations and to simultaneous differential equations.

Runge's Law
The difference of the wave-numbers of the limits of the diffuse and **Bergmann Series** in alkali spectra is equal to the wave-number of the first line of the diffuse series.

Runge's Rule
In the complex Zeeman effect, in a given magnetic field of induction B, the frequency shift of a line is either that of a Balmer line $(e/4\pi mc)B$ (*see* **Zeeman Effect**) or p/q times this, where p and q are small integers.

Russell–Saunders (Coupling) Terms
In the many-electron atom, it is convenient to classify an atomic state in terms of the total orbital angular momentum L and the total spin S. This is done in the Russell–Saunders system by assuming strong

coupling between individual spins and strong coupling between electrons to give an overall orbital momentum in the form

$$\mathbf{M}_L = \mathbf{m}_{\ell_1} + \mathbf{m}_{\ell_2} + \mathbf{m}_{\ell_3} + \ldots \mathbf{m}_{\ell_n}$$

$$\mathbf{M}_S = \mathbf{m}_{s_1} + \mathbf{m}_{s_2} + \mathbf{m}_{s_3} + \ldots \mathbf{m}_{s_n}$$

where capital letters refer to the whole atom and small letters to individual electrons. A Russell–Saunders term is written ^{2S+1}L where the superscript $2S+1$ gives the number of different $\mathbf{M}_S$ values of any state, often referred to as spin multiplicity.

In other atoms, there is a strong coupling between orbital angular momentum and spin for individual electrons, and the total angular momentum is found by summing the $\mathbf{l}+\mathbf{s}$ (i.e. giving $\mathbf{j}$) values for each electron. This interaction is known as **j–j** coupling.

Rutgers Equation
In the theory of superconduction

$$\left(\frac{\mathrm{d}H_T}{\mathrm{d}T}\right)_{T=T_C} = \sqrt{\left[\frac{1}{\mu_0 V T_C}(C^{\mathrm{s}} - C^{\mathrm{n}})_{T=T_C}\right]}$$

when SI units are used. H_T is the magnetic field intensity at which the superconductor loses its superconductivity. T_C is the temperature at which metal becomes superconducting. C^{s} and C^{n} are the specific heats in the superconducting and normal states.

Rutherford *See* **Appendix**

Rutherford Scattering
Scattering, by heavy nuclei, of light charged particles. The probability of scattering into a solid angle lying between ω and $\omega + \mathrm{d}\omega$ is given by the differential scattering cross-section $\mathrm{d}\sigma$:

$$\mathrm{d}\sigma = \frac{b^2}{16 \sin^4 \theta/2} \mathrm{d}\omega$$

where θ is the angle through which the incident particle is scattered and b is the distance of closest approach. If Ze is the charge on the nucleus, $Z'e$ and v the charge and velocity respectively of the incident particle, and μ is the reduced mass, then in SI:

$$b = \frac{ZZ'e^2}{4\pi\varepsilon_0 \mu v^2}.$$

Rydberg *See* **Appendix**

Rydberg Constant *See* **Appendix, Ritz Combination Principle** *and* **Bohr's Theory**

Rydberg–Schuster Law
The difference between the wave-numbers of the limit of the principal series and the common limit of the diffuse and sharp series, in alkali spectra, is equal to the wave-number of the first line of the principal series.

S

Sabatier–Senderens Reduction
When a mixture of carbon monoxide or carbon dioxide and hydrogen is passed over finely divided nickel at 300°C, methane is formed. Many other organic compounds may be reduced in a like manner, and all these reductions are known by this name.

Sabin *See* **Appendix**

Sabine Formula
In acoustics, a formula giving the reverberation time of an enclosure:

$$T = \frac{V}{20 \sum_{r=1}^{n} \alpha_r S_r}$$

where T is the reverberation time in seconds, V is the volume of the enclosure in cubic feet and α_1, α_2, α_3, . . . are the absorption coefficients of areas S_1, S_2, S_3, ...

Sachse–Mohr Theory

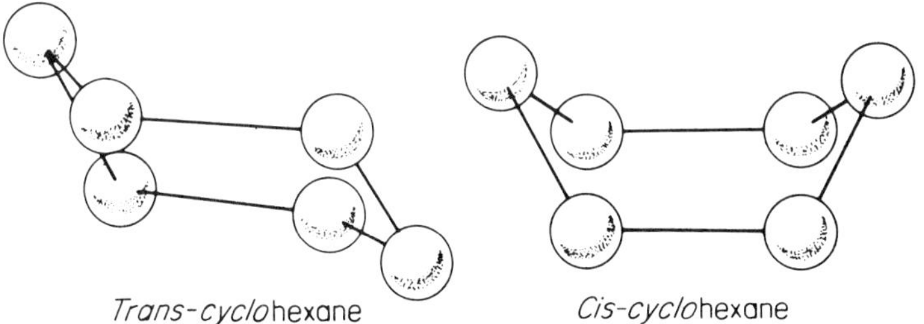

Sachse–Mohr Theory

The stability of carbocyclic rings appears to increase up to five- and six-membered rings and then to remain effectively constant. This fact can be explained if these rings are assumed to be non-planar, thus assuming that rings with six or more carbon rings are puckered. Thus, according to Sachse, *cyclo*hexane exists in two forms as shown in the figure.

Sackur–Tetrode Equation

This equation allows the translational entropy, a quantity indentical with total molar energy for a monatomic gas, to be calculated. If S_{tr} is the total molar entropy of a gas consisting of N molecules in a box of volume V where m is the mass of each molecule, then

$$S_{tr} = R\left\{ \ln\left(\frac{(2\pi m k_B T)^{3/2}}{h^3 N} \right) V + \frac{5}{2} \right\}$$

Saha's Equation

An equation giving the degree of thermal ionization ε_e of a monatomic gas

$$\log \frac{\varepsilon_e^2}{1-\varepsilon_e^2} P = -5040 \frac{E}{T} + \frac{5}{2} \log T + \log \frac{\omega_i \omega_e}{\omega_a} - 6{\cdot}491$$

where ω_i, ω_e and ω_a are constants that refer respectively to ion, electron and atom. P is the pressure in atmospheres and E is the ionization potential in volts.

Saint Venant Plasticity

Ideal plasticity, in which plastic deformation proceeds at constant yield stress.

Saint Venant's Principle

If one applies, to a small part of the surface of a body, a set of forces which are statistically equivalent to zero, then this system of forces will not noticeably affect parts of the body lying away from the above region.

Sandmeyer Reaction

When a diazonium salt is treated with a solution of a cuprous halide in the corresponding halogen acid, the diazo group is replaced by a halogen.

Sarrus' Rule

A rule for determining signs of terms in the expansion of a

determinant of order three. Write down the determinant and then repeat the first and second columns, whereby the signs of the terms in the expansion are given by the scheme:

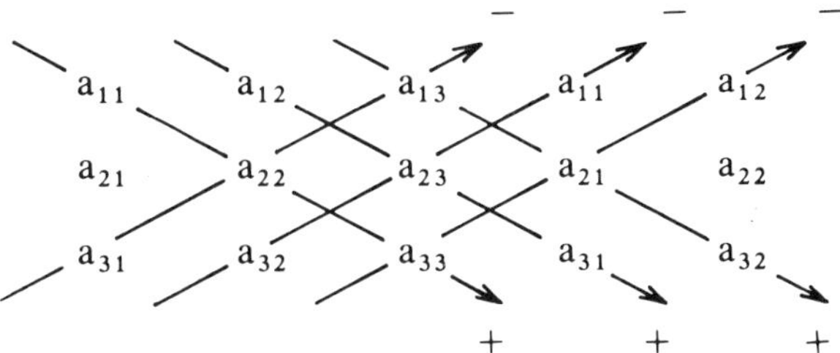

For example, term $a_{11}a_{22}a_{33}$ is positive whereas term $a_{32}a_{23}a_{11}$ is negative.

Savart *See* **Appendix**

Schiemann Reaction

The controlled thermal decomposition of dry aromatic diazonium fluoborates to give an aromatic fluoride, boron trifluoride and nitrogen.

Schiff's Bases

Aniline can be condensed with aromatic aldehydes to form anils (Schiff's bases).

$$C_6H_5-NH_2 + O=CH-C_6H_5 \longrightarrow C_6H_5-N=CH-C_6H_5$$

These bases are readily hydrolysed to the free amine and may therefore be used to protect an amine group during other processes, e.g. nitration. On reduction the anils give secondary amines, and they have been used to prepare such compounds

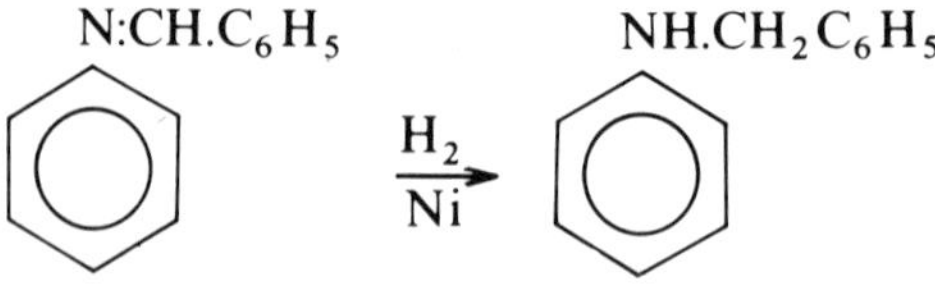

Schiff Test for Aldehydes
Rosaniline is dissolved in water and sulphur dioxide is passed through the solution until it is decolorized. Aldehydes cause the magenta colour to return to this solution.

Schlafli's Integral

$$P_n(x) = \frac{1}{2\pi j}\oint \frac{(t^2-1)^n \, dt}{2^n (t-x)^{n+1}}$$

where $P_n(x)$ is the **Legendre Coefficient** of order n. Integration is anticlockwise around a contour C encircling z in the complex plane.

Schlomilch's Infinite Product *See* **Weierstrass' Infinite Product**

Schmid's Law of Critical Stress Shear
Slip in a material takes place along a given slip plane and direction when the shear stress acting along them reaches a critical value.

Schmidt Lines
T. Schmidt (1937) observed a general regularity in the magnetic moments of nuclei. He assumed that odd-A nuclei are composed of a core forming a closed shell plus one nucleon (A is the atomic number), that the closed shell must contain an even number of neutrons and an even number of protons, and that therefore the shell has no angular momentum or magnetic moment. The nucleon which is extra to the shell has orbital angular momentum ℓ and spin s which are added vectorially to give a total angular momentum equal to the total angular momentum I of the nucleus. The Schmidt lines are lines of $\ell \pm \frac{1}{2}$ on a graph of nuclear moment against nuclear spin, and these lines form boundaries within which the nuclear moments should lie. Some deviation occurs in practice.

Schmidt Number
A dimensionless number used in diffusion and given by $(\eta/\rho)\, D_{\mathrm{AB}}$

where D_{AB} is the mass diffusivity of species A in species B, η is the dynamic viscosity and ρ the density.

Schmidt Rearrangement
Degradation of carbonyl compounds, by addition of sodium azide or of hydrazoic acid in benzene to a solution of the carbonyl compound in sulphuric acid, to yield the amine product.

Schmitt Synthesis *See* **Kolbe Synthesis**

Schoenflies Crystallographic Notation
Schoenflies treated the crystallographic symmetries as mathematical 'groups' of operations. An operation and its powers are called a *cyclic group*; thus a fourfold axis is represented by the operations:

$1, \pi/2, \pi, 3\pi/2$, i.e. powers 0, 1, 2, 3 of the fundamental rotation $\pi/2$. The symbol for a cyclic group for a rotation axis of order n is C_n. Other operations are defined by various other symbols:

Dihedral groups (a set of twofold axes at right angles to a major axis of order n) by D_n.

Octahedral groups (several axes orientated along the rational directions of a cube) by O.

Tetrahedral groups (the symmetries are those of both the octahedron and the tetrahedron) by T.

The eleven axial symmetries may be represented by the symbols C_1, $C_2, C_3, C_4, C_6, D_2, D_3, D_4, D_6, O$ and T. Other groups can be formed by adding inversion centres and reflection planes. Inversion centres are represented by subscript i and reflection planes by subscripts v, h and d dependent on whether the plane is vertical, horizontal or diagonal.

The International and Schoenflies symbols are given in the table below.

The symbol $S_4 \equiv \bar{4}$ is the only group which cannot be described by adding inversion centres or minor planes to the original eleven symbols.

In the International Symbols (**Hermann–Mauguin Symbols**), 1, 2, 3, 4 and 6 represent 1-, 2-, 3-, 4- and 6-fold rotation axes, m represents a mirror plane of symmetry and $\bar{1}, \bar{2}, \bar{3}, \bar{4}$ and $\bar{6}$ are axes of rotation-inversion.

Crystal System	Schoenflies Notation	International Symbol	Crystal System	Schoenflies Notation	International Symbol
Triclinic	C_1	1	Trigonal	C_3	3
	C_i	$\bar{1}$		D_3	32
				C_{3i}	$\bar{3}$
Monoclinic	C_2	2		C_{3v}	3m
	C_{1h}	m		D_{3d}	$\bar{3}\frac{2}{m}$
	C_{2h}	$\frac{2}{m}$			
			Hexagonal	C_{3h}	$\bar{6}$
				D_{3h}	$\bar{6}2\,m$
Orthorhombic	D_2	222		C_6	6
	C_{2v}	2mm		D_6	622
	D_{2h}	$\frac{2}{m}\frac{2}{m}\frac{2}{m}$		C_{6v}	6mm
				C_{6h}	$\frac{6}{m}$
Tetragonal	C_4	4			
	D_4	422		D_{6h}	$\frac{6}{m}\frac{2}{m}\frac{2}{m}$
	S_4	$\bar{4}$			
	C_{4h}	$\frac{4}{m}$	Cubic	T	23
				O	432
	C_{4v}	4mm			
	D_{2d}	$\bar{4}2m$		T_h	$\frac{2}{m}\bar{3}$ (m3)
	D_{4h}	$\frac{4}{m}\frac{2}{m}\frac{2}{m}$		T_d	$\bar{4}3m$
				O_h	$\frac{4}{m}\bar{3}\frac{2}{m}$, m 3 m

Schotten–Baumann Reaction

Compounds containing an active hydrogen atom can be benzoylated by means of benzoyl chloride in the presence of dilute caustic soda.

Schottky Anomaly

In a system having two (or more) energy levels such that excitation can occur between these energy levels, the variation of specific heat with temperature shows a maximum corresponding to the temperature region where there is strongest excitation to the higher energy level. The behaviour, referred to as the Schottky anomaly, has been noted particularly in paramagnetic salts.

Schottky Defect *See* **Frenkel Defect**

Schottky Effect
In thermionic emission, if the emitter temperature is kept constant, the emission current will increase as the applied field is increased. The increase is due to a lowering of the potential barrier and hence a change in the work function in the **Richardson–Dushman Equation** for the current. If current is plotted logarithmically against the square root of the electric field, a straight line is obtained called the **Schottky line**.

Schrödinger's Equation
In Schrödinger's formulation of quantum mechanics, a complex quantity called the wave-function is associated with each dynamical system. This quantity is a function of the three space coordinates and time. The dynamical properties of the system are closely related to the properties of the wave-function and the dynamical behaviour can be obtained when the wave-function Ψ of the system is known. For a one-particle system, Ψ is related to the energy by the Schrödinger equation, which may be written

$$-\frac{\hbar^2}{2m}\nabla^2\Psi + V(x,y,z)\Psi = \mathrm{j}\hbar\frac{\partial\Psi}{\partial t}$$

where m is the mass of the particle, V is the potential energy and ∇^2 is the operator

$$\frac{\partial^2}{\partial x^2} + \frac{\partial^2}{\partial y^2} + \frac{\partial^2}{\partial z^2}.$$

Usually it is the allowed energy values for stationary states which are of interest. Using **Planck's Law**, $E = h\nu = 2\pi\hbar\nu$, and putting

$$\Psi = \psi(x, y, z)\exp(-2\pi\mathrm{j}\nu t) = \psi(x, y, z)\exp(-iEt/\hbar)$$

where $\psi(x, y, z)$ is a function of position only, gives the probability density $\Psi\Psi^*$ as independent of time, although the wave-function Ψ is time-dependent. Schrödinger's equation can be written then in the time-independent form

$$\nabla^2\psi(x, y, z)+\frac{2m}{\hbar}[E-V(x, y, z)]\psi(x, y, z)=0.$$

Schur's Lemma

A reducible representation of a group is one in which all the matrices A, B, C, . . . can be transformed by a unitary matrix M such that $M^{-1}AM$ etc. have the same block diagonal form; in this all the elements are zero except for those in a set of squares along the main (leading) diagonal. Then, Schur's lemma states that any matrix which commutes with all the matrices of an irreducible representation of a group is a constant matrix, i.e. a multiple of the unit matrix.

Schwarz–Christoffel Transformation

A conformal transformation of the inside of a polygon into the upper half of a complex plane.

Schwarz Principle of Reflection

If $f(x)$ is analytic within a region D intersected by the real axis and is real on the real axis, then, for conjugate values of z, conjugate values of f are obtained, i.e.

$$f(z^*)=f^*(z)$$

where * denotes the conjugate quantity.

Schwarzschild Geometry

The space–time geometry applicable in the universe close to massive gravitational bodies. The escape velocity for bodies close to a large point mass will be a function of distance from the mass (being larger for smaller distances) and the distance at which the escape velocity approaches that of light is called the **Schwarzschild radius**. This is also the critical radius at which a massive body becomes a black hole.

Schwarz's Inequality

If f and g are any two functions, then

$$\int|f|^2\,d\tau\int|g|^2\,d\tau\geqslant\left|\int f^*g\,d\tau\right|^2$$

where f* is the complex conjugate of f, and integration is over all space. The equality only applies if one function is a scalar multiple of the other: i.e. if $g = \lambda f$ where λ is a constant.

The equivalent theorem in three-dimensional vector space is

$$(\mathbf{v}_1 \cdot \mathbf{v}_1)(\mathbf{v}_2 \cdot \mathbf{v}_2) \geqslant (\mathbf{v}_1 \cdot \mathbf{v}_2)^2.$$

(*See also* **Cauchy's Inequality**.)

Secchi's Classification

Father Secchi (1818–78) divided about 4000 stars into four groups. This represented the earliest classification of stars into spectral types.

Seebeck Effect

If a closed electrical circuit is composed of two dissimilar metals, a current flows round the circuit if one of the junctions is maintained at a different temperature from the other. The direction and magnitude of the current depend on the nature of the metals as well as on the temperatures of the junctions. An absolute measure of the effect (often called the thermoelectric power) for one metal can be obtained if the other metal is superconducting.

Segrè Chart

A chart showing the charge on nuclei (the proton number) plotted against the number of neutrons contained in the nuclei.

Seidel Aberrations

The primary aberrations found in lens systems. Deviations of light rays from the paths prescribed by **Gauss' Optics Formulae** can be expressed in terms of five sums called **Seidel sums**. These must all be zero for no monochromatic aberration, coma, astigmatism, field curvature and distortion, and it is not possible to eliminate all those aberrations at one time.

Seiffert's Spherical Spiral

A curve described on a sphere. If a sphere is taken with centre at the origin and radius unity, and if the cylindrical polar coordinates of any point on it be (ρ, ϕ, z) so that the arc of a curve traced on the sphere be given by

$$(\mathrm{d}s)^2 = \rho^2(\mathrm{d}\phi)^2 + (1-\rho^2)^{-1}(\mathrm{d}\rho)^2,$$

Seiffert's spiral is defined by

$$\phi = ks$$

where s is the arc measured from the pole of the sphere, and k is a positive constant, less than unity.

Seignette Electricity
Now known as ferroelectricity, the property of certain materials to acquire spontaneous polarization below their **Curie Point**. The phenomenon was first observed on Seignette or Rochelle salt which was first made as a laxative by the pharmacist Seignette (1672) living in La Rochelle.

Sellmeyer Formula *See* **Ketteler–Helmholtz Formula**

Serber Potential
A distribution of potential applicable to neutron–proton scattering at low energies arising from the presence of **Majorana** and **Wigner Forces**.

Serret–Frenet Formulae
If $\delta\theta$ is the small angle between tangents at neighbouring points P and Q on a curve, **t** is the unit vector along the tangent to the curve at P, **n** the unit vector in the principal normal plane given by $\mathbf{n} = \mathrm{d}\mathbf{t}/\mathrm{d}\theta$ for $\delta\theta \to 0$, and **b** is the unit vector orthogonal to **t** and **n**, making a right-handed triad of unit vectors in that order, then

$$\mathbf{t}' = \kappa\mathbf{n}$$

$$\mathbf{b}' = \tau\mathbf{n}$$

and

$$\mathbf{n}' = \kappa\mathbf{t} + \tau\mathbf{b}$$

where κ is the curvature and τ is the angle turned through per unit length of curve at the point P on the curve.

Seyfert Galaxies
A class of galaxies which are sources of exceptionally intense radiation including radiation at optical wavelengths. 1–2 per cent of all galaxies are of this type.

Shannon's Sampling Theorem
A function $x(t)$ having a power spectrum which is restricted to a finite frequency band is uniquely determined by a discrete set of sample

values. Thus, if x(t) is restricted to the frequency interval $-B \leqslant \nu \leqslant B$, then

$$x(t) = \sum_{k=-\infty}^{\infty} x\left(\frac{k}{2B}\right)\frac{\sin(2Bt-k)}{\pi(2Bt-k)} \quad (-\infty < t < \infty).$$

Sheppard's Correction

A correction factor used in statistics when calculating standard deviation σ for data which are grouped into classes of width c. If $\mu^2 = \Sigma f_s(x_s - m)^2/\Sigma f_s$ (*see* **Charlier's Checks** for the notation) and the class width is c about x_s, then the value of μ^2 and hence σ^2 should be reduced by $c^2/12$ to allow for the class width.

Shockley Partial Dislocation

The strain energy of a dislocation is proportional to b^2, where b is the magnitude of **Burgers' Vector**. Because of the consequent reduction in strain energy, some dislocations split into two partials having smaller **b** than the complete dislocation. If the partial has **b** lying in the fault plane, it is a Shockley partial dislocation (also called a 'glissile' dislocation because it is able to glide); if **b** is not parallel to the fault plane, it is a **Frank Partial Dislocation** (also called a 'sissile' dislocation because it can only diffuse and not glide).

Shubnikov–de Haas Effect

Oscillations in the magnetoresistance or **Hall Effect** in a metal or semiconductor as a function of a strong magnetic field, arising from the quantization of the energy of the electrons. (*See also the* **de Haas–van Alphen Effect**.)

Shubnikov Groups

Alternatively called magnetic groups or black-and-white groups, they are point groups of crystals in which a magnetic moment is present. In two dimensions the point group 4 mm produces two magnetic point groups 4′ m′m and 4 m′m′ where black and white shading are equivalent to magnetic moments in opposite directions;

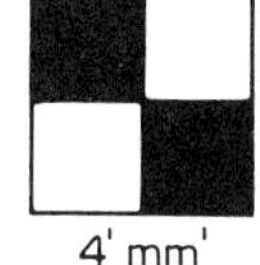

Shubnikov Groups

In three dimensions, the 32 ordinary groups (for their tabulation see **Schoenflies Crystallographic Notation**) become 58 black-and-white point groups. Similarly, 230 ordinary space groups become 1191 black-and-white space groups.

Siegbahn X Unit *See* **Appendix**

Siemens *See* **Appendix**

Silsbee Effect

Kamerlingh Onnes (1913) found that if an electric current is passed down a superconductor, then when it reaches a critical value, it destroys the superconductivity. Silsbee (1916) showed that the transition back to the normal state depends on the magnetic field associated with the current. The critical value of the magnetic field required to destroy the superconductivity is a function of temperature.

Simpson's Rule

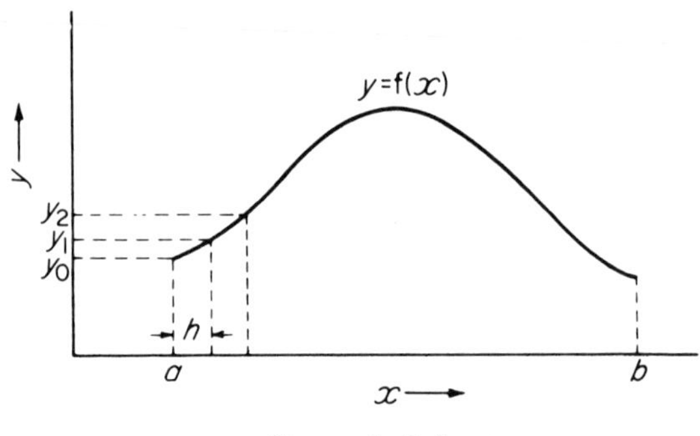

Simpson's Rule

A method of computing definite integrals. Let

$$A = \int_a^b f(x)\,dx$$

be interpreted as the area between the x axis and the curve $y = f(x)$ bounded by the ordinates $x = a$ and $x = b$. Divide the interval a–b into n equal parts, where n is even. Call $(b-a)/n = h$ and $y_0, y_1, \ldots$

the values of f(x) at a, $a+h$, . . . ; then

$$A=\frac{h}{3}(y_0+4y_1+2y_2+4y_3+2y_4+\ \dots\ +4y_{n-1}+y_n).$$

Simson's Line
The feet of the perpendiculars drawn from any point on the circumcircle of a triangle to the three sides lie in a straight line, known as Simson's line.

Skraup Synthesis
Quinoline may be prepared by heating together aniline, nitrobenzene, glycerol, concentrated sulphuric acid and ferrous sulphate. Nitrobenzene acts as an oxidizing agent and ferrous sulphate makes the reaction less violent. Arsenic acid may be used instead of nitrobenzene.

Snell's Law
For light passing from one isotropic medium to another, the ratio of the sine of the angle of incidence to the sine of the angle of refraction is a constant for a particular medium. The ratio equals the ratio of the refractive indices for the two media.

Snoek's Law
Alloys with an integral number of Bohr magnetons per atom show very little, if any, magnetostriction.

Soddy–Fajans Displacement Law
The emission of an α-particle from an atom produces a displacement to the left by two places in the periodic table, and the emission of a β-particle produces a displacement to the right by one place; i.e. the atomic numbers are reduced by 2 and increased by 1 respectively. This law is the basis of the radioactive series.

Solvay Process (for the Preparation of Sodium Carbonate)
The Solvay or ammonia–soda process depends upon the comparatively small solubility of sodium bicarbonate. The reaction occurs in three stages.

(1) Brine is saturated with ammonia gas and carbon dioxide. The solubility product of sodium bicarbonate is low and so it is precipitated.

K

(2) The sodium bicarbonate is filtered off and calcined.

$$2\,NaHCO_3 \rightarrow Na_2CO_3 + H_2O + CO_2$$

(3) The ammonia is regenerated from the mother liquor by the action of lime.

Sommelet Reaction
Aryl aldehydes may be prepared by the reaction of benzyl halides and hexamethylene tetramine by heating the reactants together in alcohol (*see* **Delépine Reaction**).

Sommerfeld Fine Structure Constant *See* **Appendix**

Sommerfeld's Theory (of the Thermal Conductivity of Metals)
A theory of thermal conductivity depending upon the free-electron theory of metals. It was assumed that the electrons in the metal behave as a degenerate gas at ordinary temperatures, and the distribution of velocities among the electrons is not given by the **Maxwell–Boltzmann Distribution Law** but by **Fermi–Dirac Statistics.**

Sørensen pH Scale
The hydrogen ion concentration of a solution can vary from 1 gram-atom per litre for strong acid to 10^{-14} gram-atoms per litre for a strong alkali. Sørensen suggested a more convenient way of expressing these quantities to avoid using negative powers of ten. He suggested that it should be expressed as pH, where

$$\mathrm{pH} = -\log a_{\mathrm{H}+}$$

where $a_{\mathrm{H}+}$ is the activity of the hydrogen ion which, for dilute solutions, may be replaced by the concentration.

Sørensen Titration
Acids such as glycine, because of their amphoteric nature, cannot be titrated directly with alkali. When formalin solution is added, a methylene amino-acid is formed.

$$H_2N.CH_2COOH + HCHO \rightarrow CH_2.N.CH_2.COOH + H_2O$$

This is a strong acid in which the amine group is protected and hence may be titrated with sodium hydroxide.

Soret Effect
If a thermal gradient is applied to a mixture or solution, a migration of atoms occurs by thermal diffusion, so that a concentration gradient is set up. This is known as the Soret effect. The converse process, whereby concentration gradients produce non-uniformity of temperature, is termed the **Dufour Effect**.

Stahl's Phlogiston Theory
All combustible bodies contain an elemental substance, phlogiston, which could be driven out by heat. The ash or residue left after combustion represented the original matter less its phlogiston.

Standt's Theorem
Every Bernoulli number B_{2n} is equal to an integer diminished by the sum of the reciprocals of all (and only those) prime numbers which, when diminished by unity, are divisors of $2n$.

Stanton's Number
The inverse of **Prandtl's Number**.

Stark Effect
Stark, in 1913, observed an effect, analogous to the transverse **Zeeman Effect**, on the hydrogen spectrum when hydrogen atoms were subjected, while radiating, to an electric field. He also observed a longitudinal effect.

In the transverse Stark effect, components exist polarized both parallel and at right angles to the field. The position and polarization of these components have been accounted for by a theory which constitutes, perhaps, one of the greatest successes of the quantum theory.

Stark–Einstein Law
Each molecule taking part in a chemical reaction induced by exposure to light absorbs one quantum of radiation causing the reaction.

Stefan's Constant *See* **Appendix**

Stefan's Law *See* **Boltzmann Law of Radiation**

Steiner Problem
A description of the general problem of linking n points, separated in

space, by a minimum path. The general problem has not been solved analytically but analogue solutions, for example using soap films, have been obtained.

Steiner's Theorem

The moment of inertia of a body depends not only on the total mass and its distribution but also upon the axis of rotation. In particular, it changes when the axis is displaced parallel to itself. Let I_c be the moment of inertia of a body of total mass M about an axis through its centre of gravity; let I_a be the moment of inertia about another axis parallel to the first and at a distance a from it. Then Steiner's theorem gives a simple relation between these quantities.

$$I_a = I_c + Ma^2$$

Steinmetz's Law

The energy dissipated per unit volume per cycle during a magnetic hysteresis loop is given by

$$W = \eta\, B_{max}^{1 \cdot 6}$$

where B_{max} is the maximum magnetic induction obtained during the cycle and η is **Steinmetz's coefficient**, which is a constant for a given material. The law holds only within a limited range of induction and, in certain materials, there is a slight deviation from the power of 1·6.

Stepanov Effect

An observed transient elèctrical current when ionic crystals are deformed. It is thought to arise from electrical charges associated with dislocations in the crystals.

Stephen Reaction

Aromatic nitrites can be reduced by stannous chloride and hydrogen chloride to imino-chlorostannates which hydrolyse easily to give the aldehyde.

Stern–Gerlach Effect

If a beam of randomly orientated atoms is passed through an inhomogeneous field, then the beam is split into a number of beams, this number depending on the kind of atom and its state. The effect shows the quantization of magnetic moment since, classically, the beam should be deflected into a vertical line.

Stevens Rearrangement
The rearrangement of organic dibenzylammonium halides on treatment with phenyl lithium or other proton acceptors. An ylide (benzylamide) is produced as an intermediate in the reaction, but this rearranges intramolecularly to give dimethylaminodibenzyl.

Stewart and Kirchhoff's Law
The emissive power divided by the absorption coefficient, for any substance, depends only on the frequency and plane of polarization of the radiation and the temperature, and is independent of the nature of the substance.

Stieltjes Integral
Let an interval (a, b) be divided into sub-intervals in the following manner

$$a = x_0 < x_1 < x_2 < \ldots < x_n = b$$

and let Δ equal the largest of these sub-intervals. Then the Stieltjes integral of $\mathrm{f}(x)$ with respect to $\alpha(x)$ from a to b is

$$\int_a^b \mathrm{f}(x)\,\mathrm{d}\alpha(x) = \lim_{\Delta \to 0} \sum_{k=1}^{n} \mathrm{f}(\zeta_k)\,[\alpha(x_k) - \alpha(x_{k-1})]$$

where $x_{k-1} \leqslant \zeta_k \leqslant x_k$.

Stiles–Crawford Effect
Light entering the eye near the centre of the pupil is usually more effective in producing a visual response than light entering near the perimeter. The effect is attributed to a variation of response of the receptors of the retina with direction of the light.

Stirling Cycle
A reversible heat cycle as used in the Stirling engine. The cycle commences with an isothermal expansion at a temperature T_1, during which heat Q_1 is taken in. This is followed by an isochoric (constant volume) fall of temperature to T_2. Next comes an isothermal compression at T_2 and, finally, an isochoric rise of temperature back to T_1.

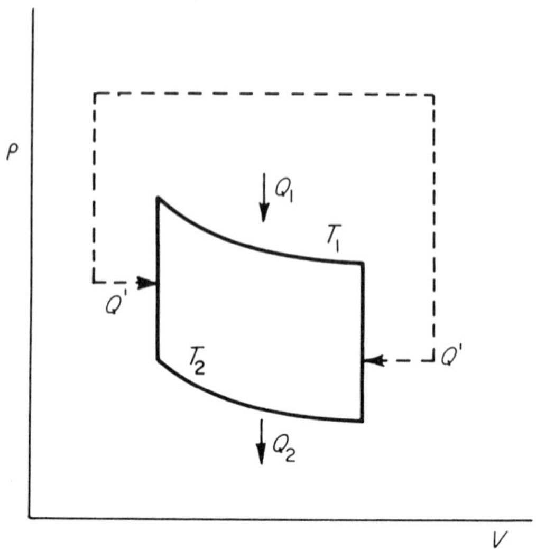

Stirling Cycle

In practice the Stirling engine has two pistons, the working piston and the recuperator piston. During the temperature fall, the gas is passed through the recuperator which absorbs heat Q'. This heat is stored while the gas expands and is returned during the final isochoric stage.

Stirling's Formula
Valid for large values of n.

$$\ln n! = n \ln n - n + \tfrac{1}{2}\ln n + \tfrac{1}{2}\ln 2\pi \ldots$$

It is often only necessary to use the first two terms.

Stobbe Condensation
The condensation between a succinic ester and carbonyl compound in the presence of sodium ethoxide. The yield may be increased by using potassium *t*-butoxide in *t*-butanol as the condensing agent. The condensation product may be refluxed with hydrobromic acid in glacial acetic acid and γ-lactone formed, catalytically reduced via the sodium salt to the acid.

Stockbarger Method *See* **Bridgman–Stockbarger Method**

Stockmayer Potential
The total potential energy Φ of two polar (ionic) molecules is of the form

$$\Phi = \frac{B}{r^{12}} - \frac{C}{r^6} - \frac{\mu^2}{r^3}\{2\cos\theta_1\cos\theta_2 - \sin\theta_1\sin\theta_2\cos(\phi_2 - \phi_1)\}.$$

The first two terms are as for the **Lennard-Jones Potential**, whereas the third term represents the interaction between two ideal dipoles described by angles θ_1 and ϕ_1 and θ_2 and ϕ_2; μ is the dipole moment of a single molecule and r the separation. The angles θ_i and ϕ_i are the usual spherical polar angles of the dipoles in a coordinate system in which the z axis is the intermolecular axis.

Stokes *See* **Appendix**

Stokes' Law
If the resistance to motion is R in a fluid of viscosity η for a particle of radius r moving at a velocity v, then

$$R = 6\pi\eta r v.$$

Stokes' Law of Radiation
It was supposed by G. Stokes that, in fluorescence, only radiations of wavelength longer than that of the exciting light could be emitted. In line spectra these are known as **Stokes' lines**. The difference in wavelength between the emitted and exciting light is called the **Stokes shift**.

Emitted lines of shorter wavelength than the incident light have also been observed (for example, in the Raman spectra of benzene), and are called **Anti-Stokes' lines**. These arise from an interaction involving lattice vibrations.

Stokes' Phenomenon
An apparent discontinuity in the value of an asymptotic expansion of a function as its argument changes its phase.

Stokes' Theorem
The surface integral of the curl of a vector over any surface is equal to

the line integral of the vector round the boundary of the surface.

$$\iint \text{curl}\, \mathbf{A}.\text{d}\mathbf{S} = \int \mathbf{A}.\text{d}\mathbf{s}$$

Stormer Unit *See* **Appendix**

Strecker Degradation
An aromatic compound which contains a carbonyl group in conjunction with another carbonyl group or nitro group will react with an α-amino acid to give the aldehyde and carbon dioxide. Measurement of the volume of carbon dioxide evolved may be used to estimate quantitatively the α-amino acid.

Strecker Reaction
When an alkyl halide is heated with sodium sulphate, the sodium salt of the sulphonic acid is produced.

Strouhal Number
A dimensionless number concerning non-steady flow of a viscous fluid past a solid body. If v is the velocity of flow of the fluid, τ a characteristic time interval for the flow (related to the rate of change) and l a characteristic length for the body, then the Strouhal number is given by $S = v\tau/l$. *See also* **Reynolds Number**.

Struve Functions
The Struve function of the mth order is

$$S_m(z) = \frac{2\,(z/2)^m}{\Gamma\,(m+\frac{1}{2})\,\Gamma(\frac{1}{2})}\int_0^{\pi/2} \sin\,(z\cos u)\sin^{2m} u\,.\,\text{d}u.$$

Sturm–Liouville Equation

$$\frac{\text{d}}{\text{d}z}\left(p(z)\frac{\text{d}y}{\text{d}z}\right) + [q(z) + \lambda r(z)]y = 0$$

where λ is the eigenvalue.

Sturm's Functions
Let $\text{f}(x)$ be a polynomial and $\text{f}'(x)$ its derivative. When $\text{f}(x)$ is divided

by $f'(x)$, let the remainder be $-f_2(x)$. Similarly, if $f'(x)$ is divided by $f_2(x)$, the remainder is called $-f_3(x)$. Thus,

$$f(x) = g_1(x)f'(x) - f_2(x)$$

$$f_1(x) = g_2(x)f_2(x) - f_3(x) \quad [f_1(x) = f'(x)]$$

$$\dots \quad \dots \quad \dots \quad \dots$$

$$f_{n-2}(x) = g_{n-1}(x)\ f_{n-1}(x) - f_n(x).$$

The degrees of the functions $f(x)$, $f'(x)$, $f_2(x)$, ... $f_n(x)$, Sturm's functions, steadily decrease, and the final remainder, $-f_n(x)$, is a constant which is only zero if $f(x) = 0$ has a multiple root.

Sturm's Theorem
If $f(x)$ is a polynomial and a and b are any real numbers ($a < b$), the number of distinct roots of $f(x) = 0$ which lie between a and b (any multiple roots which may exist being counted only once) is equal to the excess of the number of changes of sign in the sequence of **Sturm's Functions**

$$f(x),\ f'(x),\ f_2(x),\ \dots\ f_n(x)$$

where $x = a$, over the number of changes of sign in the sequence when $x = b$.

Sutherland's Formula
The viscosity η of a gas at a Kelvin temperature T is given by

$$\eta = \eta_0 \left(\frac{T}{273{\cdot}1}\right)^{3/2} \frac{C + 273{\cdot}1}{C + T}$$

where η_0 is the viscosity at 0°C and C is a constant for a given gas.

Svedberg *See* **Appendix**

Swarts' Reaction
Alkyl fluorides may be prepared by heating organic halides with inorganic fluorides such as AsF_3, SbF_3, AgF or Hg_2F_2.

Sylow's Theorem
If h is a power of a prime and divides the order of a group G, then

group G has a subgroup of order h. For example, a group of order 60 has subgroups of 2, 3, 4 and 5.

Sylvester's Theorem

In matrix theory, if the n latent roots (eigenvalues), $\lambda_1, \lambda_2, \ldots \lambda_n$ of the square matrix A are all distinct and $\mathrm{P}(A)$ is any polynomial of A,

$$\mathrm{P}(A) = \sum_{r=1}^{n} \mathrm{P}(\lambda_r) Z_r$$

where the matrix Z_r is

$$Z_r = \frac{\prod_{s \neq r} (A - \lambda_s U)}{\prod_{s \neq r} (\lambda_r - \lambda_s)}$$

and U is the unit matrix.

T

Tafel Equation

During electrolysis, if ζ is the polarization or overpotential of an electrode with respect to the electrolyte and I is the current flowing, then

$$\zeta = \frac{2{\cdot}303\, RT}{\alpha F}(\mp \log I_0 + \log I)$$

where α is the fraction of ζ assisting the dissolution of the electrode and I_0 is the exchange current or the current per unit area of the electrode at the reversible potential when $\zeta = 0$.

When ζ is plotted as ordinate against $\log I$, the curve is known as the **Tafel line**.

Talbot *See* Appendix

Talbot Bands

If a thin glass plate is placed across half the aperture of a prism so that it is in front of the *thinner* half of the prism, or if the plate is placed similarly across half the width of a grating, then a series of dark bands called Talbot bands appears in the spectrum if white light is incident on the prism or grating. These bands appear because of the path difference introduced between light passing through the two sides of the prism or grating. There is an optimum thickness of glass plate to produce maximum visibility and, if the thickness is above a critical value, the bands disappear because retardation of the light waves becomes too great for interference.

Talbot's Law

A light flashing at a frequency greater than approximately 10 hertz appears steady to the eye due to the persistence of vision. By Talbot's Law the light has an apparent intensity I given by

$$I = I_0\left(\frac{t}{t_0}\right)$$

where I_0 is the actual intensity of the light source for exposure time t during total time t_0. *See also* **Blondel–Rey Law**.

Tammann Temperature
An approximate minimum temperature at which two solids will react together at an appreciable rate.

Tate's Law
A drop falls from a narrow tube when the weight of the drop is equal to the surface tension which holds it to the liquid: $mg = 2\pi\gamma r$, where m is the mass of a drop of radius r and γ is the surface tension.

Taylor–Orowan Dislocation
An alternative name for an edge dislocation in a crystal. This can be considered as the presence in the crystal of an extra half-plane of atoms; the coordination of the extra atoms in the dislocation is less than that for the ideal lattice. *See also* **Burgers' Circuit**.

Taylor's Series

$$\mathrm{f}(x+h) = \mathrm{f}(x) + \frac{h}{1!}\mathrm{f}'(x) + \frac{h^2}{2!}\mathrm{f}''(x) + \ldots + \frac{h^n}{n!}\mathrm{f}^{(n)}(x) + R_n.$$

R_n, the remainder, is given as

$$R_n = \mathrm{f}^{(n+1)}(x+\theta h)\frac{h^{n+1}}{(n+1)!}; \quad (0 < \theta < 1) \quad \text{Lagrange's form.}$$

$$R_n = \mathrm{f}^{(n+1)}(x+\theta h)\frac{h^{n+1}(1-\theta)^n}{n!}; \quad (0 < \theta < 1) \quad \text{Cauchy's form.}$$

Taylor's series for functions of many variables is:

$$\mathrm{f}(x_1+h_1, x_2+h_2, \ldots) = \mathrm{f}(x_1, x_2, \ldots) + h_1\frac{\partial \mathrm{f}}{\partial x_1} + h_2\frac{\partial \mathrm{f}}{\partial x_2} + \ldots$$
$$+ \frac{h_1^{\,2}}{2!}\frac{\partial^2 \mathrm{f}}{\partial x_1^{\,2}} + \frac{2}{2!}h_1 h_2\frac{\partial^2 \mathrm{f}}{\partial x_1 \partial x_2} + \frac{h_2^{\,2}}{2!}\frac{\partial^2 \mathrm{f}}{\partial x_2^{\,2}} + \ldots$$

Tempkin Isotherm
A similar adsorption isotherm to the **Langmuir Adsorption Isotherm**,

but based on the assumption that the adsorption enthalpy changes linearly with pressure. It takes the form

$$\frac{V}{V_m} = c_1 \ln (c_2 kp)$$

where c_1 and c_2 are constants and the remaining notation is as for the Langmuir adsorption isotherm.

Tesla *See* **Appendix**

Thevenin's Theorem

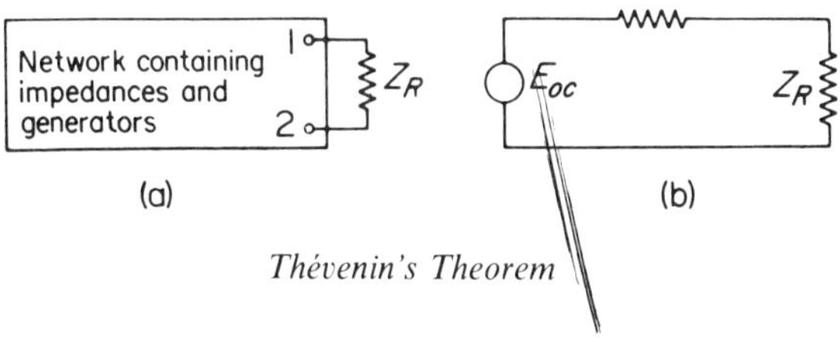

Thévenin's Theorem

The current in any impedance Z_R, connected to two terminals of a network, is the same as if Z_R were connected to a simple generator, whose generated voltage is the open-circuited voltage at the terminals in question and whose impedance is the impedance of the network looking back from the terminals, with all generators replaced by impedances equal to the internal impedances of these generators.

It has been stated that this theorem was originally enunciated by Helmholtz and should, more correctly, be called **Helmholtz's theorem**.

Thiele Benzene Formula *See* **Kekulé Benzene Formula**

Thiele's Theory of Partial Valencies
In order to explain abnormal 1 : 4 addition as well as 1 : 2 addition of halogens to conjugated double bonds in olefins, Thiele postulated the existence of a symmetrical bond system in these compounds. Instead of alternate double and single bonds, he envisaged that all bonds had the character of a 1·5 bond. This left equal possibility of the addition of the halogen to any carbon atom, hence explaining the abnormal addition observed experimentally. The modern explanation depends on the existence of resonance hybrids.

Thomas–Fermi Potential
A coulombic potential in which there is screening of the test charge by a surrounding electron gas. *See also* **Coulomb's Law of Force**.

Thom's Classification Theorem
In catastrophe theory, a theorem which classifies the number of possible elementary catastrophes which can occur depending on the dimension of the quantity required to describe the cause C.

Dimensions of C	1	2	3	4	5	6
Number of elementary catastrophes	1	2	5	7	11	(∞)

For example, for one dimension only, the fold catastrophe occurs, but for four dimensions or factors in the cause, seven elementary catastrophes arise called fold, cusp, swallowtail, butterfly, hyperbolic, elliptic and parabolic. The cusp catastrophe which occurs for dimensions of C of two or more is also called the **Riemann–Hugoniot Catastrophe** because of application to shock waves.

Thomsen–Berthelot Principle
J. Thomsen (1854) and M. Berthelot (1867) suggested that the heat evolved in a chemical reaction was a measure of the 'affinity' of the reacting substances. This view, which implies that only exothermic reactions can occur spontaneously, must be incorrect, as for each exothermic reaction there must be an endothermic reaction.

Thomson Effect
Heat is absorbed or liberated when an electric current passes through a single homogeneous but unequally heated conductor. The coefficient of the Thomson effect for a given metal is defined as the heat absorbed per second when a current of 1 ampere flows from one point in the metal to another whose temperature is 1°C higher, over and above the heat developed according to **Joule's Law**. Also called the **Kelvin effect**.

Thomson–Freundlich Equation
The solubility of component B in a solid solution of α is given by

$$\ln\frac{[B_r^{\alpha}]}{[B_{r\to\infty}^{\alpha}]} = \frac{2M\sigma}{rRT\rho}$$

where $[B_r^{\alpha}]$ is the concentration of B at saturation for particles of radius r, and $[B_{r\to\infty}^{\alpha}]$ is the concentration for large particles; M is the atomic weight and ρ the density of component B. σ is the interfacial free energy or surface tension.

Thomson Parabolas

Thomson showed the existence of isotopes for various elements by passing ions of the elements through crossed magnetic and electric fields and on to a photographic plate, which then showed up a pattern of parabolas called Thomson parabolas. Each parabola corresponds to a particular mass/charge ratio and hence to a particular isotope.

Thomson Scattering

The scattering, with no change of wavelength, of electromagnetic radiation by electrons. Thomson showed the scattering cross-section to be independent of frequency and to have a classical value of $6{\cdot}65 \times 10^{-29}\,\mathrm{m}^2$. This type of scattering is unimportant for radiation of visible wavelengths but is important at x-ray wavelengths. *See also* **Thomson Cross-section (Appendix).**

Thomson's Formula

The period of oscillation of the current during the discharge of a capacitor is given by

$$T = \frac{2\pi}{\sqrt{(1/LC - R^2/4L^2)}}$$

where C is the capacitance, and L and R are the associated series inductance and resistance.

Thomson's Theoretical Gas/Liquid Curve

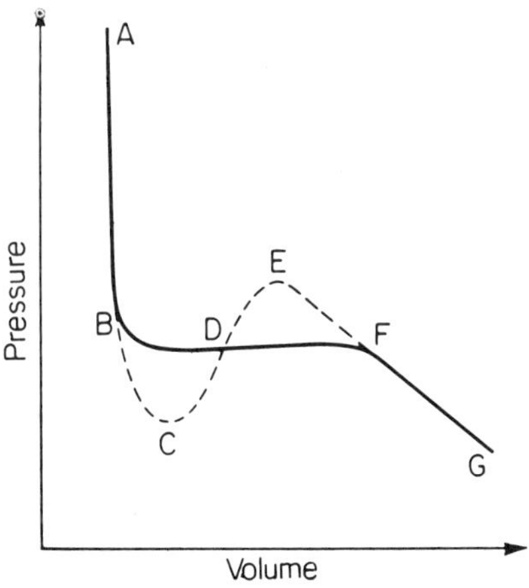

Thomson's Theoretical Gas/Liquid Curve

p/V curves for real gases follow the curve ABFG. The suggestion was made by J. J. Thomson that the ideal behaviour of a gas should be represented by the curve ABCDEFG, thus emphasizing the essential continuity between the liquid and gas phases. This postulate may have no theoretical significance but it enables the p/V relation to be expressed as a simple mathematical equation, whereas it is difficult to express discontinuous behaviour of the type represented by ADG.

Thomson (Thermocouple) Relations

For two conductors A and B connected at their two ends, with a temperature difference ΔT between the junctions, the Peltier heat is given by

$$\frac{d\Sigma}{dT} = \frac{\Pi_{AB}}{T}$$

where Σ is the Seebeck voltage between the junctions Π_{AB} is the Peltier coefficient for a junction between metals A and B.

The Thomson heat is given by

$$\mu_A - \mu_B = -T\frac{d^2\Sigma}{dT^2}$$

where μ_A and μ_B are the Thomson coefficients for conductors A and B. (*See also* **Peltier, Seebeck** *and* **Thomson Effects**.)

Thue's Theorem

The equation

$$a_0x^n + a_1x^{n-1}y + \ldots + a_ny^n = C$$

where $n \geqslant 3$ and C is integral, and where the form on the left is irreducible, cannot have an infinite number of integer solutions x, y.

Tischenko Reaction

All aldehydes can be made to undergo the **Cannizzaro Reaction** in the presence of aluminium ethoxide. Under these conditions the ester is formed.

Titius–Bode Law *See* **Bode's Law**

Tollens Reagent

A solution of silver nitrate in excess ammonium hydroxide solution oxidizes aldehydes to the corresponding acids with reduction of the complex silver ion to a characteristic mirror of elemental silver. It is important that the excess of ammonia is only just sufficient to form the complex argentiammonium ion, otherwise the silver mirror is redissolved by the solution.

Torr *See* **Appendix**

Torricellian Vacuum
The space enclosed above a column of mercury when a sealed tube is filled with mercury and inverted so that its lower end is inserted in a volume of mercury. The height of the mercury column will then be limited to that which the atmosphere can support. The enclosed space is evacuated except for mercury vapour.

Torricelli's Law of Efflux
The velocity of efflux of a liquid through an orifice is equal to that which a body would attain in falling freely from the free surface of the liquid to the orifice.

Townsend Coefficient
The number of ionizing collisions per centimetre of path of a charged particle in the direction of the applied electric field.

Townsend Discharge
A discharge in which additional ionization is solely due to ionization of the gas by electron collisions.

Traube's Rule
Traube (1891) found that, for dilute solutions, the concentration of a member of a homologous series at which equal lowering of surface tension was observed decreased threefold for each additional methylene group in any given series.

Troland *See* **Appendix**

Trouton's Law
The molar latent heat of vaporization of a liquid, divided by its boiling point at atmospheric pressure on the absolute scale of temperature, is equal to approximately 23 if the latent heat is expressed in calories (or

85 if the latent heat is expressed in joules). This generalization was first discovered by A. Pictet (1876) and rediscovered by W. Ramsay (1877) and F. Trouton (1884).

Tschebyscheff *See* **Chebyshev**

Tschirnhausen's Cubic *See* **Catalan's Trisectrix**

Twaddell (Twaddle) Scale
A scale of specific gravity, where degrees Twaddell are given by

$$^{\circ}\mathrm{Tw} = (\mathrm{sp.gr.} - 1) \times 200$$

Tyndall Effect
If a strong beam of light is passed through a true solution, the path of the beam cannot be seen. If the solution, however, contains colloidal particles, the beam becomes visible when viewed at right angles to its path, by virtue of the scattering of light by the particles. The ultramicroscope depends on the effect. (*See also* **Rayleigh Scattering**.)

U

Ullmann Reaction
Triphenylamine may be prepared by heating together diphenylamine, iodobenzene, potassium carbonate and a little copper powder in nitrobenzene.

Urbach's Rule
Optical absorption in many semiconductors increases exponentially with energy near the absorption edge, with a proportionality coefficient in the exponent of $\sigma/k_B T$, where σ is a constant, approximately 1.

V

Van Allen Radiation Belts

Two toroidal-shaped regions surrounding the earth, with their axes coinciding with the earth's geometric axis and containing circulating particles of low energy but high intensity. Particles in the outer belt probably originate from the sun, those in the inner belt from the radioactive decay of neutrons liberated in the atmosphere by cosmic radiation.

Vandermonde Determinant

The determinant (or its transpose) having

(i) unity for each element of the first row,
(ii) an unspecified second row,
(iii) the elements of the ith row as the $(i-1)$th power of the corresponding elements of the second row.

Vandermonde's Theorem

If m and n are any numbers whatsoever

$$(m+n)_r = m_r + C_1{}^r m_{r-1} n_1 + C_2{}^r m_{r-2} n_2 + \dots + n_r$$

where

$$n_r = n(n-1)(n-2)\dots(n-r+1)$$

and

$$C_k^r = \frac{r!}{(r-k)!k!}$$

Van der Waals Adsorption

Two main types of adsorption have been distinguished, depending upon whether the association between the gas and solid is physical or chemical in nature. Physical adosorption depends for cohesion between gas and solid on the so-called Van der Waals forces, which are the normal cohesive forces between any molecules. It is this type of adsorption which is known by the above name. In the other case the adsorption depends upon valence forces. Van der Waals adsorption is characterized by relatively low heats of adsorption, of the same order as the heat of vaporization of a gas. The equilibrium between gas and solid is reversible and is attained rapidly when the pressure and temperature are changed.

Van der Waals Equation
An equation of state for a real gas, which may be expressed in the form

$$\left(p+\frac{a}{V^2}\right)(V-b)=RT.$$

a/V^2 is a measure of the attractive force of the molecules and is called the cohesive pressure, whilst b is the covolume which is equal to four times the actual volume of the molecules.

Van der Waals–London Interaction
The dipole interaction between two neutral atoms, given by

$$\Phi_r = -(C/r^6)$$

where C is a constant, r is the separation of the atoms and Φ_r is the potential energy of the interaction. (*See also* **Lennard-Jones Potential.**)

Van Hove Singularities
Divergencies in the slope, with respect to energy E, of the electron or phonon density of states in a crystalline material; i.e. divergencies in $\mathrm{d}g_n/\mathrm{d}E$, where g_n is the density of states in energy level E_n. They reflect the fact that the group velocity associated with the electrons or phonons vanishes at some frequencies.

Van't Hoff Equation (for Osmotic Pressure)
Van't Hoff pointed out in 1886 the analogy between gases and solutions. He deduced, from Pfeffer's work on osmotic pressure, an equation for the osmotic pressure of a dilute solution analogous to the **Clapeyron Equation of state** for gases:

$$\pi V = RT$$

where π is the osmotic pressure, V is the dilution, T is the absolute temperature and R is a constant almost exactly equal in value to the ordinary gas constant.

The osmotic pressure of a solution of an electrolyte cannot be expressed by a simple Van't Hoff Equation. In order to make allowance for the deviation, Van't Hoff proposed a modification to the equation.

$$\pi V = iRT$$

where i is known as the **Van't Hoff factor**. If n molecules of an electrolyte are contained in a certain volume and the degree of dissociation of the molecules is α, then

$$i = \alpha(n-1)+1.$$

Van't Hoff Isochore
This equation, which is of fundamental importance in chemistry, represents the variation, with temperature, of the equilibrium constant for a reaction involving gases, in terms of the change in heat content; it may be written

$$\frac{\mathrm{d} \ln K_p}{dT} = \frac{\Delta H}{RT^2}$$

where K_p is the equilibrium constant and ΔH is the change of heat of reaction at constant pressure.

A similar equation for reaction at constant volume may be deduced.

Van't Hoff Isotherm
This equation expresses the change of free energy in terms of the temperature, the equilibrium constant and some factor expressing the concentration and the number of molecules of each reactant.

$$-\Delta G = RT \ln K_c - RT \Sigma\, v - \ln c,$$

where c is concentration in g l^{-1} and v is the number of molecules of reactants.

Van't Hoff–Le Bel Theory
In 1874 Van't Hoff and Le Bel gave, independently, a solution to the problem of optical isomerism. Van't Hoff postulated that the four bonds of a saturated carbon atom were directed to the corners of a tetrahedron. Le Bel's theory is similar in nature; he postulated only that, whatever the spatial arrangement of the bonds, the compound

Cabcd is asymmetrical. This asymmetry leads to the possibility of spatially different isomers.

Van't Hoff Principle
If the temperature of a chemical reaction in equilibrium is raised, the amount of the reactants formed by an endothermic reaction is increased, whereas lowering of the temperature increases the amount of the products produced by an exothermic reaction.

Van't Hoff Principle of Superposition
The optical rotatory power of a substance containing a number of asymmetrical carbon atoms is equal to the algebraic sum of the contributions of each carbon atom taken alone. This contribution is independent of the configuration of the rest of the molecule.

Van Vleck Paramagnetism
In considering magnetic susceptibility as a function of temperature, one limiting case is when the level splitting is $\gg k_B T$. Levels $\gg k_B T$ above the ground state make a contribution to the susceptibility which is independent of temperature. This term is known as Van Vleck paramagnetism and is in addition to the normal $1/T$ and diamagnetic terms.

Varley Effect
In an ionic solid, an energetic charged particle or photon can ionize a negative ion to give it a positive charge and induce it to take up an interstitial position. (*See also the* **Wigner Effect**.)

Vegard's Law
In an alloy system, lattice spacings show a linear dependence on composition, having values between those for the pure elements. The law was originally proposed for mutually soluble pairs of ionic salts, and exceptions to the rule are particularly numerous among metallic solid solutions.

Venn Diagram (or **Euler Diagram**)
A graphical method of illustrating **Boolean Algebra** in terms of an algebra of classes. It consists of a number of outlined areas (usually circles or other single shapes) inside an enclosing rectangle containing

the class. Each outlined area represents a variable which can be true or false; within the area it is true (represented by A, B, C, . . .) whereas outside the area but within the rectangle it is false (represented by $\bar{A}$,

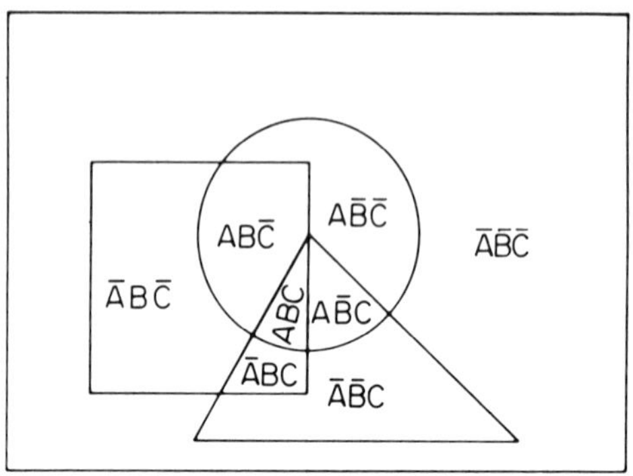

Venn Diagram

$\bar{B}$, $\bar{C}$, . . .). Where areas overlap, more than one variable is true. For instance, the enclosing rectangle could represent people. Areas enclosing A, B, C, . . . would represent those people possessing particular attributes, whereas those areas enclosing $\bar{A}$, $\bar{B}$, $\bar{C}$, . . . would represent people in whom the attributes are absent.

Verdet's Constant
The angle of rotation of the plane of polarization per unit length per unit magnetic field in the **Faraday Effect**. Verdet's Constant is approximately proportional to the square of the wavelength of the light.

Verneuil Method
A method of single-crystal growth, originally intended for the preparation of artificial gem-stones, in which powder is dropped through an oxy-hydrogen flame so that it falls molten on to a crystal seed.

Villari Effect
The susceptibility of an iron wire is increased by stretching when the magnetism is below a certain value, but diminished when above that value.

Violle *See* **Appendix**

Voigt Effect
When light is passed through a vapour (or liquid) in a direction which is perpendicular to an applied magnetic field, the vapour becomes doubly refracting. This arises because the absorption frequencies in the vapour are different for light polarized perpendicular and parallel to the magnetic field. The magnitude of the effect is proportional to the square of the magnetic field strength. *See also the* **Cotton–Mouton Effect**.

Voigt Profile
The shape of a spectral line when Doppler broadening and Lorentz damping both apply. (*See also* **Doppler Effect** *and* **Lorentzian Line Shape**.)

Volt; Volt, Thermal *See* **Appendix**

Volta Effect
The e.m.f. established when two dissimilar metals are brought into contact with each other.

Volterra Dislocation
Consider a body such that it can be cut across without being separated, i.e. an annulus of material. A cut can be made across one side of the ring and the cut ends can be displaced relative to each other and rejoined. The nature of this type of displacement has been considered by Volterra, who showed that, if the stresses in the ring are to be single-valued, the surfaces of the cut must be displaced rigidly with respect to each other. The discontinuities in the components of the displacement vector along the Cartesian axes must satisfy certain conditions.

The simplest discontinuities of this kind are shown in the figure.

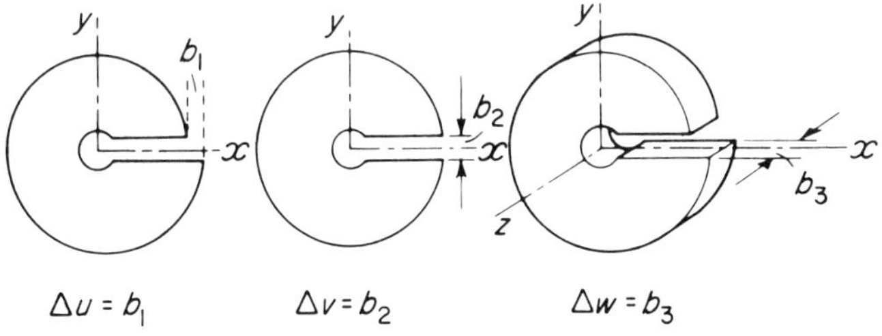

Volterra Dislocation

Volterra Equation
An integral equation of the type

$$\psi(z) = \int_a^z K(z, z_0)\psi(z_0)dz_0 + \phi(z)$$

for the unknown function ψ. $K(z, z_0)$ is known as the kernel of the equation.

Von Braun Reaction
Tertiary amines displace the bromide ion from a cyanogen bromide to give a quaternary salt. An alkyl group of this compound is attacked by the bromine ion spontaneously to give a dialkyl cyanamide.

Vorlander's Rule
If the first substituent in the benzene nucleus contains an unsaturated valency, then the group is *m*-orientating; if it is saturated, then *o–p* substitution will take place. There are exceptions to this rule, for instance cinnamic acid (directing group CH=CH.COOH) should be *m*-, whereas in fact it is *o*–*p*-orientating.

Voronoy Polyhedron
A crystal cell which is defined in a similar way to the **Wigner–Seitz Cell** but is defined for any discrete set of points. These are not necessarily lattice points forming a **Bravais Lattice**. Consequently, the structure and orientation of a Voronoy polyhedron will depend on the particular point enclosed within the array.

Wagner Rearrangement

When alcohols containing no hydrogen atoms adjacent to the alcohol group are dehydrated, dehydration and molecular rearrangement occur together.

Walden Inversion

The Walden inversion may be defined as the conversion of a laevorotatory optical isomer to the dextrorotatory isomer without recourse to resolution. The conversion may be either total or partial. The phenomenon was discovered when the following series of reactions was first carried out.

$$\begin{array}{ccccc} \text{CHOH.COOH} & \xrightarrow{PCl_5} & \text{CHCl.COOH} & \xrightarrow{AgOH} & \text{CHOH.COOH} \\ | & & | & & | \\ CH_2COOH & \xleftarrow[KOH]{} & CH_2COOH & \xleftarrow[PCl_5]{} & CH_2COOH \end{array}$$

l-malic acid | *d*-chlorosuccinic acid | *d*-malic acid

Walden's Rule

According to P. Walden (1906) in an electrolyte the product $\Lambda_0\eta_0$ is constant, where Λ_0 is the equivalent conductivity at infinite dilution and η_0 is the viscosity of the solvent. The rule breaks down as the effective diameter of the ions in different solvents varies due to different degrees of solvation.

Wallach Transformation

When gently warmed with concentrated sulphuric acid, azoxy benzene rearranges to *p*-hydroxyazobenzene.

Wallis' Theorem

$$\frac{\pi}{2} = \frac{2.2.4.4.6.6.8.8\ldots}{1.3.3.5.5.7.7.9\ldots}$$

$$= \prod_{k=1,2,\ldots} \left(\frac{2k}{2k-1}\cdot\frac{2k}{2k+1}\right)$$

Wannier Exciton *See* **Frenkel Exciton**

Wannier Function
A wave-function used to represent electrons in crystalline solids. It is the **Fourier Transform** of the eigenfunction for electron momentum. Any Wannier function can be expressed as a linear combination of **Bloch's Functions** or vice versa.

Warburg's Law
The energy liberated during a complete hysteresis cycle is $\oint H\mathrm{d}M$, where H is the magnetic field strength, M the magnetization and the relation between magnetic induction B and magnetization is taken as $B = \mu_0 H + M$.

Waring's Formula

$$\frac{1}{x-a} = \frac{1}{x} + \frac{a}{x(x+1)} + \frac{a(a+1)}{x(x+1)(x+2)} + \ldots$$

Watt *See* **Appendix**

Watt's Curve
The locus of the midpoint of a segment of a straight line whose ends move on two circles of equal radii.

Watt's Law
The latent heat of steam at any temperature of generation, added to the sensible heat required to raise the water from 0°C to the temperature, is constant. Regnault, in 1847, showed that Watt's Law was materially in error, and that the total heat of steam increased with the temperature of generation.

Weber *See* **Appendix**

Weber–Fechner Law
The intensity of a sensation is proportional to the logarithm of the physical stimulus which produces it. Seeing, hearing and photographic tone-reproduction follow the law closely.

Weber's Bessel Functions of the Second Kind *See* **Bessel Functions**

Weber's Number
A dimensionless number for flow of fluid past a body, given by

$$W = \rho v^2 l/\gamma$$

where ρ is the density of the fluid, v the velocity, γ the surface tension and l a characteristic linear dimension of the body.

Weddle's Rule
A method of computing definite integrals similar to **Simpson's Rule**. It is more accurate than Simpson's rule but n must be divisible by 6. For this rule

$$A = \frac{3h}{10}(y_0 + 5y_1 + y_2 + 6y_3 + y_4 + 5y_5 + 2y_6 + 5y_7 + y_8 + \dots$$
$$+ 2y_{12} + \dots + 2y_{n-6} + 5y_{n-5} + y_{n-4} + 6y_{n-3}$$
$$+ y_{n-2} + 5y_{n-1} + y_n).$$

(*See also* **Newton–Cotes Formula**.)

Weerman Degradation
An α-hydroxy or α-methoxy amide may be degraded to an aldehyde containing one less carbon atom by the action of a cold aqueous solution of sodium hypochlorite. This reaction has been used by Haworth to descend the sugar series.

Weierstrass' Approximation Theorem
A continuous function may be approximated to any degree of accuracy over a closed interval by a polynomial.

Weierstrass Function
A kind of elliptic function with a double pole and zero in each periodic lattice.

Thus, if

$$z = \int_{\mathscr{P}(z)}^{\infty} \frac{\mathrm{d}t}{(g_1 t^3 - g_2 t - g_3)^{1/2}}$$

then $\mathscr{P}(z)$ is the Weierstrassian. *See also* **Lamé's Functions**.

Weierstrass Inequalities
If $a_1, a_2, \ldots$ are positive numbers less than 1 whose sum is denoted by S_n, then

$$1 - S_n < (1 - a_1)(1 - a_2) \ldots (1 - a_n) < \frac{1}{1 + S_n}$$

$$1 + S_n < (1 + a_1)(1 + a_2) \ldots (1 + a_n) < \frac{1}{1 - S_n}$$

where, in the last inequality, it is supposed that $S_n < 1$.

Weierstrass' Infinite Product

$$\frac{1}{\Gamma(x)} = x\,\mathrm{e}^{\gamma x} \prod_{s=1}^{\infty} \left(1 + \frac{x}{s}\right) \mathrm{e}^{-x/s}$$

where γ denotes **Euler's Constant**.

Weierstrass' Test for Convergence
If an infinite series $a_0(x) + a_1(x) + a_2(x) \ldots$ of real or complex functions converges uniformly and absolutely on every set of values x such that $a_n(x) \leqslant M_n$ for all n, where $M_0 + M_1 + M_2 \ldots$ is a convergent comparison series of real positive terms, then the series converges uniformly.

Weissenberg Effect
The tendency of a fluid to creep up along a rotating shaft instead of being thrown out by centrifugal force.

Weisskopf Unit *See* **Appendix**

Weiss System
A system of indexing planes in terms of their intercepts on three

suitably chosen axes, such that the intercepts are expressed as integral multiples of the three basic crystal dimensions. This system has been replaced by **Miller Indices** where reciprocals are used.

Weiss Zone Law

If any face with **Miller Indices** *hkl* lies in the zone $[UVW]$ defined by faces $h_1k_1\ell_1$, $h_2k_2\ell_2$ so that $U = k_1\ell_2 - \ell_1k_2$, $V = \ell_1h_2 - h_1\ell_2$ and $W = h_1k_2 - k_1h_2$, then

$$Uh + Vk + W\ell = 0.$$

A zone is a set of faces with mutually parallel intersections, and $[UVW]$ is the zone symbol specifying the common direction of these intersections.

Weizsäcker's Formula

A semi-empirical expression for the binding energy of nuclei:

$$E(Z,A) = \underset{(1)}{a_1A} - \underset{(2)}{a_2A^{2/3}} - \underset{(3)}{a_3\frac{Z^2}{A^{1/3}}} - \underset{(4)}{\frac{a_4}{A}\left(Z-\frac{A}{2}\right)^2} + \underset{(5)}{\begin{cases} +a_5A^{-1/2} & Z \text{ even, } N \text{ even} \\ 0 & Z \text{ even, } N \text{ odd or } Z \text{ odd, } N \text{ even} \\ -a_5A^{-1/2} & Z \text{ odd, } N \text{ odd} \end{cases}}$$

where $E(Z, A)$ is the binding energy, A the atomic number, Z the number of protons and N the number of neutrons $(A = Z + N)$. The terms represent:

(1) A binding energy proportional to the volume.
(2) A correction to the volume term for surface effects.
(3) A correction for the coulombic energy of the nucleus of assumed radius $A^{1/3}$.
(4) and (5) Terms taking into account symmetry properties in the nuclear state giving a tendency for $Z = N$ and for even Z and even N.

Weldon Process

A process for making chlorine which depends on the action of manganese dioxide on hydrochloric acid. It is still used to a small extent on the industrial scale. The manganese chlorides formed are mixed with lime in tall iron towers and air is blown through for several hours until the manganese is re-oxidized to manganese dioxide, which is then allowed to settle and re-used in the first stage.

Wentzel, Kramers, Brillouin, Jeffreys Method (W. K. B. J. Method)
A short-wavelength approximation used in potential and wave-motion problems.

The solution of

$$\frac{d^2\psi_n}{dx^2} + \{k^2 - U(x)\}\psi = 0$$

can be written as

$$\psi = \exp\{\phi(x)\}$$

where $d\phi/dx$ satisfies the equation

$$\left(\frac{d\phi}{dx}\right)^2 + \frac{d^2\phi}{dx^2} + q^2 = 0$$

if $q^2 = k^2 - U(x)$ is a slowly varying function of x.

Werner's Coordination Theory
A theory which accounts for the formation of stable compounds such as the chloroplatinates from platinic chloride and hydrochloric acid, both of which have no residual valency in the usually accepted sense of the word. The atoms or radicals in the nucleus are said to be coordinated with the central atom and, since they are not ionized, must be attached by co-valencies. The existence of these compounds can now be explained quantitatively in terms of ligand field theory.

Whiddington's Law
When a solid is bombarded by primary electrons to produce secondary emission (i.e. the emission of additional lower-energy electrons), the energy loss of the primary electrons per unit path length is given by

$$-\frac{d\,E_p(x)}{dx} = \frac{A}{E_p(x)}$$

where E_p is the energy of the primaries and A is a constant which is characteristic of the solid. This leads to the result that the range of the primaries is proportional to the square of their energy.

Whitmore Mechanism

Whitmore suggested a mechanism for the pinacol–pinacolone rearrangement

$$(CH_3)_2C(OH).C(OH).(CH_3)_2 \rightarrow (CH_3)_3C.CO.CH_3 + H_2O.$$

It depended on the affinity of the two carbon atoms in the asymmetric free radical formed by removing a hydroxyl group from pinacol being greater for the methyl group on the other carbon atom than for a returning hydroxyl group. It is now only of historical interest and the rearrangement can be explained in terms of the electromeric effect.

Whitworth's Theorem

If there are N events and r possible conditions, such that every single condition is satisfied in N_1 of the events, every two of the conditions are simultaneously satisfied in N_2 of the events, . . . and finally all r conditions are simultaneously satisfied in N_r of the events, then the number of events free from all the conditions is

$$N - C_1{}^r N_1 + C_2{}^r N_2 - C_3{}^r N_3 + \ldots + (-1)^r N_r$$

where

$$C_n{}^r = \frac{r!}{(n-r)!r!}$$

Wiedemann Effect

The lower end of a vertical wire, magnetized longitudinally, twists, if free, in a certain direction when a current passes through it.

Wiedemann, Franz and Lorenz's Law

The ratio of the thermal to the electrical conductivity has the same value at the same temperature for all good conductors; for different temperatures the value of the ratio is proportional to the absolute temperature.

Wiedemann's Law

The empirical law that the susceptibility χ_L of a solution containing m grams of salt of susceptibility χ_A dissolved in M grams of solvent of susceptibility χ_B is

$$\chi_L = \frac{m\chi_A + M\chi_B}{M + m}$$

Wien Effect

If a very high electric field is applied to an electrolyte so that the ion moves through many times the diameter of the ionic atmosphere during the time of relaxation, the ion should be virtually free from the retarding effect of its atmosphere, and the value of the conductivity should approach that at infinite dilution.

Wiener and Hopf Equation

A type of integral equation which has some importance in problems of diffraction from semi-infinite obstacles. It is of the form

$$\psi(x) = \int_0^\infty \mathrm{U}(|x - x_0|)\psi(x_0)\,\mathrm{d}x_0 \quad x > 0$$

where $\mathrm{U}(|x - x_0|)$ is a given real function of the absolute difference between x and x_0, $\psi(x)$ is a given function and $\psi(x_0)$ is unknown.

Wiener–Khintchine Theorem

This relates the power spectrum of a random process to the correlation function which indicates how fast the random process is changing. If the random function $y(t)$ is stationary (this means that there is no preferred origin in time for describing y) and its correlation function is

$$\mathrm{K}(\tau) \equiv \langle y(t)\,y(t+\tau) \rangle$$

(averaging over time t) then, by expressing $y(t)$ as a **Fourier Integral**

$$\mathrm{K}(\tau) = \int_{-\infty}^{+\infty} \mathrm{J}(\omega)\mathrm{e}^{j\omega\tau}\,\mathrm{d}\omega$$

where $\mathrm{J}(\omega)$ is the 'spectral density' of y (the power density per unit frequency bandwidth averaged over the total time). Conversely

$$\mathrm{J}(\omega) = \frac{1}{2\pi}\int_{-\infty}^{+\infty} \mathrm{K}(\tau)\,\mathrm{e}^{j\omega\tau}\,\mathrm{d}\tau.$$

Wien's Constant *See* **Appendix**

Wien's Displacement Laws

The curve of energy of the emission of a black body at a given temperature against the wavelength of the emission passes through a maximum. By means of the following relationships, Wien was able to account for the change of distribution of energy of the spectrum as the temperature was raised, provided the energy curve at one temperature was known:

$$\lambda_m T = \text{constant}; \quad \frac{E_m}{T^5} = \text{constant}$$

where λ_m is the wavelength of the maximum energy E_m at absolute temperature T. *See* **Planck's Quantum Theory**.

Wierl Equation
An expression for the intensity of scattering of electrons off a molecule consisting of a number of atoms. The angular distribution of intensity is given by

$$I(\theta) = \sum_{ij} f_i f_j \frac{\sin(sR_{ij})}{sR_{ij}}.$$

θ is the scattering angle, f_i and f_j are scattering factors for atoms separated by distance R_{ij} and s is given by

$$s = (4\pi/\lambda)\sin(\theta/2)$$

where λ is the wavelength associated with the scattered electrons.

Wigner Coefficients Alternatively called **Clebsch–Gordan Coefficients**

Wigner Crystallization
The crystallization into a non-conducting state which can occur for an electron gas of low density in the presence of a background of uniform positive charge called 'jellium'. Such a crystallization is not observed in real materials, as it is not possible to produce the properties of 'jellium'.

Wigner Effect
A neutron or charged particle with large kinetic energy can, on striking an atom, send it into an interstitial position, thus producing a **Frenkel Defect.** To return such atoms to their original lattice positions it is necessary to anneal the material. The effect can occur, for instance, in graphite used as a moderator in nuclear reactors.

Wigner Force
A force between nucleons where there is no space or spin exchange. The force is always attractive. (*See also* **Majorana Force.**)

Wigner–Seitz Cell
A type of cell in a crystal lattice, obtained geometrically by taking planes which perpendicularly bisect the lines connecting a single lattice point to all surrounding lattice points. The minimum volume containing the lattice point that can be defined in this way is the Wigner–Seitz cell. It is a primitive unit cell as not only can all lattice space be filled by the cell, using suitable crystal translation operations, but it is also a minimum-volume cell for which this is possible. The Wigner–Seitz cell is therefore the region of space that is closer to a single lattice point than to any other lattice point, and may be defined as such.

Wigner's Theorem
If ψ is an eigenfunction of H, the Hamiltonian operator (*see* **Hamiltonian**), describing a quantum-mechanical system, and ψ corresponds to an eigenvalue E, and if R is one of a group of symmetry elements of H (for instance, R could be a rotational or reflectional symmetry operation), then $R\psi$ is an eigenfunction of H corresponding to the same eigenvalue E.

Wijs' Method
The iodine value, a measure of the unsaturation of an oil, may be determined by titration of the oil with iodine chloride in glacial acetic acid.

Willgerodt Reaction
In this reaction, carbonyl compounds are converted into amides containing the same number of carbon atoms. The reaction was carried out originally by heating an aryl alkyl ketone with a solution of yellow ammonium sulphide, but a modified technique is to heat the ketone with approximately equivalent amounts of sulphur and dry amine.

Williamson Synthesis
This is a method of producing ethers, particularly mixed ethers.

Sodium or potassium ethoxide is heated with an alkyl halide:

$$RONa + R'X \rightarrow ROR' + NaX$$

Williot Diagram
A graphical method of obtaining deflections at points in a structural framework under load.

Wilson's Theorem
If, and only if, p is a prime, then $(p-1)! + 1$ is divisible by p.

Wilson Transition
A transition from metallic to non-metallic behaviour, which can occur in materials where the metallic behaviour arises from band-overlap but where the overlap disappears on contraction of the lattice parameter under high pressure. *See also* **Mott Transition**.

Winslow Effect *See* **Johnsen–Rahbek Effect**

Wittig Reaction
This is a method of preparing olefins in which the position of the double bond is known exactly. Alkylidenephosphoranes react with the carbonyl compound to produce the olefin and a substituted phosphine oxide.

$$(C_6H_5)_3P{=}CR_1R_2 \rightleftharpoons (C_6H_5)_3\overset{+}{P}{-}\overset{-}{C}R_1R_2$$

$$(C_6H_5)_3\overset{+}{P}{-}\overset{-}{C}R_1R_2 + O{=}C\begin{matrix}R_3\\R_4\end{matrix} \rightarrow \begin{matrix}(C_6H_5)_3\overset{+}{P}{-}CR_1R_2\\ \overset{-}{O}{-}CR_3R_4\end{matrix} \rightarrow (C_6H_5)_3PO + \begin{matrix}R_1\\R_2\end{matrix}C{=}C\begin{matrix}R_3\\R_4\end{matrix}$$

Wittig Rearrangement
Benzyl alkyl ethers, when reacted with organolithium compounds, undergo replacement of one of the hydrogen atoms of the methylene group attached to the benzene ring by a lithium atom. The molecule then rearranges and, finally, produces the phenylalkylcarbinol.

Wohl Degradation
The aldo-sugar series may be descended by this method. For example an aldo-hexose is converted into its oxine. The reaction of acetic

anhydride with this compound dehydrates the oxine group to the cyanogen group and acetylates the hydroxy groups. When the acetyl derivative is warmed with ammoniacal silver nitrate, the acetyl groups are removed by hydrolysis and a molecule of hydrogen cyanide is eliminated, with the formation of the pentose.

Wolffenstein–Boters Reaction
When benzene is subjected to the action of mercuric nitrate in nitric acid, that is oxynitration, 2:4 dinitrophenol and picric acid are formed.

Wolff–Kishner Reduction
When the hydrazone or semicarbazone of an aldehyde or ketone is heated with potassium hydroxide or sodium ethoxide in a sealed tube, the corresponding hydrocarbon is obtained.

Wronskian
Given a second-order ordinary differential equation, its any two solutions, say y_1 and y_2, are said to be independent, i.e. not proportional to each other, when their Wronskian

$$\Delta(y_1, y_2)$$

does not vanish identically.

$$\Delta(y_1, y_2) = y_1\frac{dy_2}{dx} - y_2\frac{dy_1}{dx}$$

Wulff Net
A stereographic net on to which are projected the poles of the planes of a crystal. The net is so drawn that it can be used to measure inter-planar angles directly from the relative positions of the pole projections.

Wulff Theorem
The surface tension of a solid–fluid interface varies with the crystallographic direction of the solid surface. This variation may be represented by a surface, the **Wulff surface**, which is obtained by taking the termini of radial vectors of length proportional to the surface tension for the particular crystallographic direction. If planes are drawn perpendicular to the ends of these vectors, then the shape of

the body lying inside these planes is, by Wulff's theorem, the shape of the crystal at equilibrium in the fluid. If there are pronounced cusps in the Wulff surface, then plane faces will be present on the crystal.

Würtz–Fittig Reaction
Homologues of benzene may be prepared by warming an ethereal solution of alkyl and aryl halides with sodium.

Würtz Reaction
When an ethereal solution of an alkyl bromide or iodide is treated with sodium, the hydrocarbon is formed. It has been found that the Würtz reaction gives good yields only for paraffins containing even numbers of carbon atoms, and that it usually fails with tertiary alkyl halides.

Y

Young's Fringes

The bright and dark fringes established on a viewing screen set up in front of two narrow parallel slits in an opaque screen which is illuminated from the rear with monochromatic light. The two slits act effectively as coherent-light sources and a diffraction pattern is produced as a consequence of interference of the light from the two sources. If a monochromatic light source is not used, coloured fringes result.

Young's Modulus

The ratio of the stress to the fractional increase in length, within the limit of proportionality, for a solid subjected to a uniform tension along one axis.

Yukawa Potential

A potential function V used by Yukawa in his investigation of atomic nuclear fields. For a radially symmetric charge distribution, the potential at distance r from the centre of the charge distribution is

$$V = \frac{g}{r}\mathrm{e}^{-r/L}$$

where g and L are constants. A function of this type can be obtained as a static solution of the **Klein–Gordon Equation**.

Z

Zeeman Effect

Zeeman found that, when a beam of light passed through a magnetic field, any spectral line of frequency $p/2\pi$ was split up into

(*a*) two lines of frequency $(p \pm \Delta p)/2\pi$ for the light emitted along the lines of magnetic force (Longitudinal effect),
(*b*) three lines of frequency $p/2\pi$, $(p \pm \Delta p)/2\pi$ for the light emitted at right angles to the lines of force (Transverse effect)
where $\Delta p = cB/2mc$. B is the magnetic induction.

In (*a*) the two lines are circularly polarized with right-handed (for the line of lower frequency) and left-handed (for the line of higher frequency) polarization with respect to the magnetic field.

In (*b*) the radiation is composed of three plane polarized constituents, the polarization of the central component being perpendicular to that of the other two and parallel to the lines of magnetic force.

This is the simple Zeeman effect, theoretically explained by Lorentz using classical electron theory. More complex resolutions observed in practice (anomalous Zeeman effect) are now understood and are found to be in agreement with the quantum theory of atomic structure and radiation. (*See also* **Appendix – Zeeman Displacement.**)

Zener Effect

According to wave mechanics, there is a finite probability that an electron be found outside a potential well, even if its energy is insufficient for it to surmount the potential barrier. This effect is known as quantum-mechanical tunnelling and offers an explanation of breakdown in certain electrical materials when a field is applied. In this context it is known as the Zener effect.

For example, the reverse breakdown of a p–n junction can be attributed to the tunnelling of electrons from the full valence band of one semiconductor into the empty conduction band of the other, this explaining the mechanism of operation of the Zener diode. In this case breakdown may also occur due to avalanche multiplication, but these two mechanisms can be distinguished by a study of the temperature-dependence of the current–voltage characteristics. Dielectric breakdown of insulators may also sometimes be attributed to this effect.

Zeno Paradoxes

Zeno posed four mathematical paradoxes, the most famous being that of Achilles and the tortoise. The tortoise is given a start over Achilles. Although Achilles is ten times faster than the tortoise, by taking incremental distances which are always one tenth that between Achilles and the tortoise, Zeno argued that Achilles would never catch the tortoise. The argument is false, because decreasing times are associated with the decreasing incremental distances. By summing to infinity the distance series such that the overall distance is achieved, and is achieved in finite time, it can be shown that Achilles catches the tortoise.

Zerewitinoff Determination

A method of analysis of organic compounds containing active hydrogen atoms, as in the hydroxy, carboxy and imino groups, by reaction with methyl magnesium halide to yield methane.

Zinke Nitration

Ortho- and *para-* bromine or iodine atoms (but not fluorine or chlorine) substituted into phenols can be replaced by a nitro group by the action of nitrous acid or the action of acetic acid on a nitrate.

Zintl's Rule

In salt-like ionic compounds, only those elements which precede the noble gases by one to four places are able to become negative ions. Exceptions to the rule have been found in alloys of indium and gallium, and the rule should be modified to include elements up to five places before a noble gas.

Appendix: A list of named units

Amagat
A unit of volume often used in the study of the equation of state of gases. By definition the molar volume of a gas at 0°C and 1 atmosphere is 1 amagat. The exact value of the unit depends upon the gas considered, but is approximately 22·4 litres $mole^{-1}$. There is also an amagat unit of density equal to the density of a gas in which 1 mole occupies a volume of 1 amagat.

Ampere (A)
The ampere is the practical unit of electric current, the absolute unit in the MKS system, and one of the seven basic SI units. An absolute ampere is the steady current which, when maintained in two parallel rectilinear conductors of infinite length and negligible cross-section and separated by a distance of 1 metre in free space, produces between these conductors a force of 2×10^{-7} newton per metre length.

There is also an international ampere which is defined electrochemically, being the current which under specified conditions will deposit $1{\cdot}118 \times 10^{-6}$ kilogram of silver from a silver nitrate coulometer.

Ampere-hour (A h)
The ampere-hour is a unit of quantity of electricity. It is the amount of electricity which is passed when 1 ampere has been flowing for 1 hour, i.e. 1 A h = 3600 coulombs.

Ampere, Thermal
The SI unit of thermal current. 1 thermal ampere corresponds to an entropy flow of 1 watt per kelvin. Formerly, the thermal ampere corresponded to a heat flow rate of 1 watt.

Ampere-turn (A, sometimes At, AT)
The ampere-turn is the practical, MKS, and SI unit for magnetomotive force. 1 ampere-turn is the magnetomotive force resulting

from the passage of a current of 1 ampere through one turn of a coil.

Ångström (Å)
A unit of length equal to 10^{-10} m.

Avogadro's Number
The number of individual atoms in a gram-atom, of ions in a gram-ion, or of molecules in a gram-molecule, is called Avogadro's number. The accepted value is $6{\cdot}02252 \times 10^{23}$ per mole.

Balmer
A term sometimes used for the unit of wave-number. It is the number of waves in a centimetre and has units of cm^{-1}. *See also* **Kayser** *and* **Rydberg.**

Baud
A pulse per second or alternatively one binary digit. (Named after J. M. E. Baudot.)

Bel (B)
A number used to express logarithmically the ratio of two powers. It is named after Alexander Bell and is defined as

$$N = \log (P_2/P_1)$$

where P_1 and P_2 are the two powers. The practical unit is the decibel (dB) where 1 decibel = 0·1 bel.

Biot (Bi)
A c.g.s. unit of current. 1 biot is the constant current which, if maintained in two parallel rectilinear conductors of infinite length and negligible cross-section and with separation of 1 centimetre in a vacuum, would produce a force between the conductors equal to 2 dynes per centimetre of length. 1 biot = 10 amperes.

Blondel
A metric unit of luminance. One blondel is the luminance of a uniform diffuser emitting 1 lumen per square metre.

$$1 \text{ blondel} = \frac{1}{\pi} \text{ candela metre}^{-1}.$$

Bohr Magneton (μ_B)
A unit of magnetic moment.

$$\mu_B = \frac{eh}{4\pi m} = 9{\cdot}2741 \times 10^{-24} \text{ joule tesla}^{-1}.$$

In c.g.s. units, μ_B equals $9{\cdot}2741 \times 10^{-21}$ erg gauss^{-1}.

Bohr Radius (a_0)
The radius of the first Bohr orbit of the hydrogen atom. In SI,

$$a_0 = h^2/\pi\mu_0 c^2 m e^2 = 5{\cdot}29177 \times 10^{-11} \text{ m}.$$

Boltzmann's Constant (k)

$$\begin{aligned} k &= 1.38041 \times 10^{-23} \text{ joule deg}^{-1} \\ &= 8.6167 \times 10^{-5} \text{ eV deg}^{-1}. \end{aligned}$$

Brewster (B)
The unit used to measure the stress-optical constant when a material shows birefringence under an applied stress. A brewster is the number of angstroms per millimetre path by which one component of the light is retarded relative to the other if a stress of 1 bar (atmosphere) is applied in a direction perpendicular to the path of the light.

$$1 \text{ B} = 10^{-12} \text{ Pa}^{-1}.$$

Brig
A ratio of two units to base 10 (named after Briggs).

Clausius
1 clausius is the entropy associated with a temperature of 1 kelvin in which there is an increase in heat of 1000 calories.

$$1 \text{ clausius} = 4186{\cdot}8 \text{ joules kelvin}^{-1}.$$

Compton Wavelength (λ_C)

$$\lambda_C = \frac{h}{mc} = 2{\cdot}42631 \times 10^{-12} \text{ m} \quad \text{(for electron)}.$$

See also **Compton Effect**.

Coulomb (C)
The coulomb is the practical unit of electric charge and the absolute,

MKS and SI unit. The coulomb is the charge crossing the section of a conductor in which a steady current of 1 ampere flows for 1 second. As the coulomb is directly related to the charge on the electron, which is $1{\cdot}602 \times 10^{-19}$ coulomb, it may be considered as a fundamental electrical unit. In the c.g.s. electrostatic system, the statcoulomb is defined in terms of the electrostatic force between charges. A statcoulomb is that charge which, when placed 1 centimetre away from a like charge, repels it with a force of 1 dyne.

$$1 \text{ coulomb} = 2{\cdot}9959 \times 10^{9} \text{ statcoulombs.}$$

Coulomb, Thermal
The SI unit of thermal charge. 1 thermal coulomb corresponds to an increase in entropy of 1 joule per kelvin.

$$1 \text{ thermal coulomb} = 1 \text{ J K}^{-1}.$$

Formerly, the unit corresponded to an amount of heat of 1 joule.

Curie (Ci, formerly c)
That quantity of any radioactive substance which has a decay rate of $3{\cdot}7 \times 10^{10}$ disintegrations per second.

Dalton
The name occasionally used for atomic mass number, equal to 1/12 the mass of a neutral carbon-12 atom.

$$1 \text{ dalton} = 1{\cdot}66033 \times 10^{-27} \text{ kg.}$$

Debye (D)
A unit of electric dipole moment used for the dipole moments of molecules.

$$1 \text{ debye} = 3{\cdot}33564 \times 10^{-30} \text{ coulomb metre.}$$

Einstein
A unit used to describe the photoenergy involved in a gram-molecule of a substance during a photochemical reaction. The photoenergy E in Einstein units is given by $N_0 h\nu$ where N_0 is Avogadro's number and ν is the frequency of the electromagnetic radiation in hertz.

Eötvös (E)
The change in gravitational acceleration of 10^{-9} galileo over a

horizontal distance of 1 centimetre.

$$1\ \mathrm{E} = 10^{-9}\ \mathrm{m\ s^{-2}}\ (\text{horizontal m})^{-1}.$$

Farad (F)
1 farad is the capacitance of a condenser having a charge of 1 coulomb when the potential difference across the plates is 1 volt. The energy stored is 0·5 joule. The unit is very large and the everyday unit is the microfarad, which equals 10^{-6} F.

Faraday
The faraday is the electric charge carried by 1 gram-equivalent of an ion and is equal to 96 487 coulombs. (This value uses the international scale of atomic masses.)

Farad, Thermal
The SI unit of thermal capacitance. 1 thermal farad is the thermal capacity for which an amount of entropy of 1 joule per kelvin added to a body raises its temperature by 1 kelvin.

$$1\ \text{thermal farad} = 1\ \mathrm{J\,K^{-2}}.$$

Formerly, the thermal farad corresponded to 1 joule of heat resulting in a temperature rise of 1 kelvin. 1 thermal farad = 1 J $\mathrm{K^{-1}}$.

Fermi
An obsolete unit of length equal to 10^{-15} m. Radii of nuclei are of this order of magnitude.

Finsen Unit (FU)
A unit of intensity of ultraviolet light at a specified wavelength, usually 296·7 nm. 1 FU = 10^5 W $\mathrm{m^{-2}}$.

Fourier *See* **Ohm Thermal**

Franklin (Fr)
An electrostatic c.g.s. unit of charge. 1 franklin is that charge which exerts on an equal charge at a distance of 1 centimetre in a vacuum a force of 1 dyne. 1 Fr = $(0{\cdot}1/c)$ coulomb.

Fraunhofer (F)
A unit of reduced width of a spectrum line given by

$$W = 10^6\ \Delta\lambda/\lambda$$

where W is the reduced width in fraunhofers, λ is the wavelength and $\Delta\lambda$ is the equivalent width of the line.

Fresnel
A unit of optical frequency sometimes used in spectroscopy.

$$1 \text{ fresnel} = 10^{12} \text{ Hz}.$$

Galileo *or* **Gal** (Gal)
Unit of acceleration in the c.g.s. system of units equal to 1 cm s^{-2}.

Gauss (G)
This unit is the c.g.s. unit of magnetic flux. If a straight wire is passed through a magnetic field so as to cut it at the rate of 1 cm s^{-1} perpendicular to the direction of the induction, then the value of induction necessary to produce an electromotive force of 1 abvolt per centimetre length of the wire is 1 gauss. It is equal to 1 maxwell per square centimetre. The corresponding SI unit is the tesla:

$$1 \text{ tesla} = 10^4 \text{ gauss}.$$

Gibbs
A unit of adsorption. 1 gibbs is an adsorption of 10^{-6} mole over 1 square metre.

Gilbert (Gb)
The c.g.s. electromagnetic unit of magnetomotive force. 1 gilbert is the magnetomotive force resulting from the passage of a current of 4π abamperes through one turn of a coil.

$$1 \text{ gilbert} = 10/4\pi \text{ ampere-turns}.$$

Hartree
A unit of energy used in atomic studies and equal to me^4/h^2. 1 hartree $\approx 4{\cdot}8505 \times 10^{-18}$ J. *See also* **Hartree System of Units**.

Hefnerkerze (HK) (**Hefner**)
The unit of luminous intensity used in Germany until 1940. 1 Hefnerkerze $\approx 0{\cdot}901$ international standard candle.

Helmholtz
A metric unit of dipole moment per unit area.

$$1 \text{ helmholtz} = 3{\cdot}33564 \times 10^{10} \text{ C m}^{-1}.$$

Henry (H)
The henry is the practical, MKS and SI unit of inductance. It is that inductance in which an induced electromotive force of 1 volt is produced when the inducing current is changed at the rate of 1 ampere per second. In practice the millihenry is of convenient magnitude.

Henry, Thermal
The SI unit of thermal inductance. 1 thermal henry is the thermal inductance for which an entropy flow of 1 watt per kelvin is associated with a kinetic energy of 1 joule. 1 thermal henry = $1 \text{ J K}^2 \text{ W}^{-2}$.

Formerly the thermal henry corresponded to a heat flow of 1 watt associated with a kinetic energy of 1 joule.

$$1 \text{ thermal henry} = 1 \text{ J W}^{-2}.$$

Hertz (Hz)
Unit of frequency equal to cycles per second.

Joule (J)
This is the fundamental unit of energy. It is the work done when the point of application of a force of 1 newton is displaced through a distance of 1 metre in the direction of the force.

Kapp Line
A unit of magnetic flux equal to 6000 maxwells.

Kayser (K)
The accepted name for the unit of wave-number, cm^{-1}.

$$1 \text{ K} = 100 \text{ m}^{-1}.$$

Kelvin (K, formerly °K)
The SI unit of thermodynamic temperature or temperature difference as defined from the **Kelvin Temperature Scale**.

Historically the unit kelvin has also been used for the kilowatt-hour.

Lambert (L)
In illumination, a perfectly diffusing surface is one for which the luminous intensity per unit area in any direction varies as the cosine of the angle between the direction and the normal to the surface. It appears therefore to be equally bright, no matter in which direction it is viewed. The lambert is a unit of brightness defined in terms of such a surface and is the luminance of a uniform diffuser which emits a total flux of 1 lumen per square centimetre.

$$1\ \mathrm{L} = 1/\pi \text{ candela cm}^{-2}.$$

If the diffuser emits a flux of 1 lumen per square foot, then it has a luminance of 1 **foot-lambert**.

Langley
A measure of radiant energy received per unit area.

$$1 \text{ langley} = 1 \text{ calorie cm}^{-2}.$$

Lorentz
The **Zeeman Displacement** ($e/4\pi mc$) expressed in wave-numbers.

$$1 \text{ Lorentz unit} = 46{\cdot}689\ \mathrm{T}^{-1}\ \mathrm{m}^{-1}.$$

Loschmidt Number
The Loschmidt number is the number of molecules of a gas in a volume of 1 millilitre at 0°C and 1 atmosphere of pressure. The calculation of its value using the kinetic theory of gases, which attributes to each molecule a spherical collision cross-section, leads to a nearly constant value for this quantity, for simple gases, of about $2{\cdot}7 \times 10^{19}$. For gases of more complex molecules the Loschmidt number may vary from this value, but is usually of the order of 10^{19}.

Maxwell (Mx)
The c.g.s. electromagnetic unit of magnetic flux. 1 maxwell is the magnetic flux which, linking a circuit of one turn, produces in the circuit an electromotive force of 1 abvolt as it is reduced to zero at a uniform rate in 1 second. $1\ \mathrm{Mx} = 1\ \mathrm{abV\ s} = 10^{-8}\ \mathrm{Wb}$. (*See also* **Weber**.)

Mayer
A metric unit of specific heat. 1 mayer is the quantity of heat in joules

required to raise the temperature of 1 gram of a substance by 1 kelvin.

$$1 \text{ mayer} = 10^3 \text{ J kg}^{-1} \text{ K}^{-1}.$$

McLeod
A unit of pressure level. The pressure level P in mcleods is related to the pressure p in millimetres of mercury by $P = \log(1/p)$. 1 mcleod is the pressure level corresponding to 10^{-1} mm Hg.

Neper (Np)
A unit used to express the scalar ratio of two currents, voltages or amplitudes. Named after John Napier (Latin: *Ioanne Neper*), it can be thought of as a measure of attenuation. It is defined as

$$N = \ln(a_2/a_1)$$

where a_1 and a_2 are two amplitudes. 1 neper corresponds to an amplitude ratio of e.

Newton (N)
A unit of force in the SI and MKS systems. 1 newton is that force which induces in 1 kilogram an acceleration of 1 metre per second per second.

Oersted (Oe)
A c.g.s. electromagnetic unit of field strength. In a magnetic field of strength 1 oersted, unit magnetic moment experiences a couple of 1 centimetre. 1 Oe $= 10^3/4\pi$ A m^{-1}.

Ohm (Ω)
The SI, MKS and practical unit of resistance, reactance and impedance. 1 ohm is the resistance between two points of a conductor when a constant potential difference of 1 volt applied between these two points produces in this conductor a current of 1 ampere, the conductor not being a source of any electromotive force.

Ohm, Acoustical (Ω_a)
A c.g.s. unit of acoustical impedance. 1 acoustical ohm is the ratio of the sound pressure level of 1 dyne per square centimetre to a source sound strength of 1 cubic centimetre per second.

$$1\ \Omega_a = 10^5 \text{ Pa s m}^{-3}.$$

There is also the **specific acoustical ohm** (Ω_s) which is a c.g.s. measure of specific impedance. 1 specific acoustical ohm is the ratio of the sound pressure level of 1 dyne per square centimetre to a sound particle velocity of 1 centimetre per second. $\Omega_s = 10$ Pa s m^{-1}.

Ohm, Mechanical (Ω_m)
A c.g.s. unit of mechanical impedance. 1 mechanical ohm is the ratio of a force of 1 dyne to a velocity of 1 centimetre per second.

$$1\ \Omega_m = 10^{-3}\ \mathrm{N\ s\ m^{-1}}.$$

Ohm, Thermal
An SI unit of thermal resistance. 1 thermal ohm is the thermal resistance for which a temperature difference of 1 kelvin causes an entropy flow of 1 watt per kelvin. 1 thermal ohm = 1 K^2 W^{-1}.

The unit is also called the **fourier**. It was formerly the thermal resistance that allowed a heat flow of 1 watt under a temperature difference of 1 kelvin.

Pascal (Pa)
The SI unit of pressure. 1 pascal is the pressure resulting from a force of 1 newton acting uniformly over an area of 1 square metre.

$$1\ \mathrm{Pa} = 1\ \mathrm{N\ m^{-2}}.$$

Planck
An SI unit of action. It is the action of an energy of 1 joule over 1 second.

Planck's Constant

$$h = 6{\cdot}6262 \times 10^{-34}\ \mathrm{J\ s}.$$

Poise (P)
A unit of dynamic viscosity. If the shearing stress per unit area, τ, in a liquid is expressed in dynes and the velocity gradient perpendicular to the direction of flow is $\mathrm{d}u/\mathrm{d}z$, then the viscosity η defined by $\eta = \tau/(\mathrm{d}u/\mathrm{d}z)$ is in units of poise. 1 P = 0·1 N s m^{-2}.

Poncelet
A metric unit of power. It is the power available when a force which is able to accelerate 100 kilograms by 1 metre per second per

second is displaced through a distance of 1 metre in the direction of the force in 1 second.

$$1 \text{ poncelet} = 980{\cdot}655 \text{ watts.}$$

Preece

A unit of electrical resistivity used historically for describing the resistivity of insulators. 1 preece = $10^{13}\,\Omega\,\text{m}$.

Ray

A unit of mechanical or acoustical impedance named after Lord Rayleigh and equal to 1 acoustical or mechanical ohm.

Rayl

A unit of specific acoustical impedance named after Lord Rayleigh. 1 rayl is the ratio of a sound pressure of 1 dyne per square centimetre to a sound particle velocity of 1 centimetre per second.

$$1 \text{ rayl} = 10 \text{ Pa s m}^{-1}\,.$$

Rayleigh (R)

A unit of luminous intensity used especially for the night sky. $1\text{ R} \approx 5{\cdot}272 \times 10^{-25} f$ watt metre^{-2} steradian^{-1} where f is frequency in hertz.

Reyn

A unit of dynamic viscosity. 1 reyn is the dynamic viscosity that gives rise to a tangential stress of 1 poundal per square foot across two planes separated by 1 foot when the streamlined flow is 1 foot per second. 1 reyn = $1{\cdot}4882\text{ N s m}^{-2}$.

Röntgen (R, formerly r)

A measure of the ionizing effect of γ or x-rays in air. A röntgen is the dose of electromagnetic radiation which will produce in air a charge of $2{\cdot}58 \times 10^{-4}$ coulomb on all the ions of one sign, when all the electrons of both signs liberated in a volume of air of mass 1 kilogram are completely stopped.

Rutherford (rd)

That quantity of any radioactive substance which has a decay rate of 10^6 disintegrations per second. (*See also* **Curie**.)

Rydberg (Energy)
1 rydberg of energy $= e^2/a_0 \approx 2{\cdot}4252 \times 10^{-18}$ J. ($a_0 = h^2/me^2$.) (*See also* **Hartree System of Units**.)

Rydberg (Reciprocal length)
Reciprocal length of a distance of 1 centimetre. 1 rydberg $= 100\ \mathrm{m}^{-1}$. *See also* **Balmer**, **Kayser**, *and* **Rydberg Constant**.

Rydberg Constant
A fundamental physical constant of value 10 973 731 metre^{-1}. It is the numerical value of $\mu_0{}^2me^4c^3/8h^3$ and is used particularly in atomic physics. *See also* **Rydberg**.

Sabin
A unit of absorption in acoustics. If V is the volume in cubic feet of an enclosure and T the reverberation time in seconds, the absorption a in sabins is given by

$$a = 0{\cdot}161\ V/T.$$

Savart (s)
A measure of pitch interval. If I_s is the pitch interval in savarts between two frequencies f_1 and f_2, then

$$I_s = 1000 \log (f_2/f_1).$$

Thus an octave ($f_2/f_1 = 2$) is divided into 301 savarts. There is also a modified savart given by

$$I_{ms} = \frac{300}{\log 2} \log (f_2/f_1).$$

Siegbahn X Unit (X, also Xu)
A unit of wavelength defined by the fact that, at 18°C, the spacing of the cleavage planes of calcite is given by 3029·45 X units.

$$1\ \mathrm{X} = 1{\cdot}00202 \times 10^{-13}\ \mathrm{m}.$$

Siemens (S)
The SI unit of conductance. 1 siemens is the conductance between two points of a conductor when a difference of potential of 1 volt applied between these two points produces in this conductor a current of

1 ampere, the conductor not being a source of any electromotive force. The unit is also called the **mho** (reciprocal ohm).

Sommerfeld Fine Structure Constant

$$\alpha = \mu_0 e^2 c/2h = 7{\cdot}297351 \times 10^{-3} \approx 1/137$$

A constant appearing in the relativistic correction to the energy of the electron orbits in the atom.

Stefan's Constant

$$\sigma = 2\pi^5 k_B{}^4/15\, h^3 c^2 = 5{\cdot}6696 \times 10^{-8}\ \mathrm{W\ m^{-2}\ K^{-4}}.$$

See also **Boltzmann's Law of Radiation.**

Stokes (St, formerly S)
A c.g.s. unit of kinematic viscosity. When the dynamic viscosity η is measured in poise and the density ρ is measured in grams per cubic centimetre, the kinematic viscosity (η/ρ) is in stokes.

$$1\ \mathrm{St} = 10^{-4}\ \mathrm{m\ s^{-1}}.$$

Stormer Unit (S)
The trajectories of charged particles in the field of a magnetic dipole become independent of the charge Ze, momentum p and mass m of the particles if all the lengths are expressed in Stormer units given (using c.g.s.–e.m.u.) by

$$S = \left(\frac{Zem}{pc}\right)^{1/2}.$$

The unit has particular application to charged nuclei entering the earth's magnetic field.

Svedberg (S)
A rate of sedimentation of 10^{-13} centimetre per second for unit acceleration during centrifuging; i.e. it is the rate of sedimentation under a force of 1 dyne. It is defined from the equation $S = 10^{-3}\, v/\omega^2 x$ where v is the velocity of the boundary between the solution containing the molecules and the solvent, x is the distance from the axis of rotation and ω is the angular velocity in radians per second.

Talbot
A unit of luminous energy. 1 talbot is the luminous energy corresponding to 1 joule of radiant energy for a luminous efficiency of 1 lumen per watt. 1 talbot = 1 lumen second.

Tesla (T)
The SI unit of magnetic flux density.

$$1 \text{ tesla} = 1 \text{ weber metre}^{-2} = 10^4 \text{ gauss}.$$

Thomson Cross-section

$$\frac{8\pi r_0{}^2}{3} = 6{\cdot}65 \times 10^{-28} \text{ m}^2.$$

r_0 is the effective electron radius given by $\mu_0 e^2/4\pi m$. (*See also* **Thomson Scattering**.)

Torr
Named after Torricelli, it is the pressure of 133·322 pascals. A standard atmosphere has a pressure of 760 torr.

Troland
The retinal illumination produced by a surface having a luminance of 1 candela per square metre when the area of aperture of the eye is 1 square millimetre. The unit is also called a luxon.

Violle
The luminous intensity of 1 square centimetre of platinum at its solidification temperature (2045 K). The candela (cd) is the present unit of luminous intensity and 1 violle = 20·17 candelas.

Volt (V)
The SI unit of electric potential, potential difference or electromotive force, named after Count Volta. 1 volt is the electric potential difference between two points in a current-carrying conductor carrying a constant current of 1 ampere when the power dissipated between the two points is 1 watt.

Volt, Thermal
A thermal potential difference across a conductor corresponding to a temperature difference of 1 kelvin.

Watt (W)
The SI unit of power. 1 watt is the power available when 1 joule of energy is expanded in 1 second.

Weber (Wb)
The SI unit of magnetic flux. It is the magnetic flux which, linking a circuit of one turn, produces in it an electromotive force of 1 volt as it is reduced to zero at a uniform rate in 1 second. It has also been used in the c.g.s.–e.m.u. system for a magnetic pole strength of 1 dyne per oersted.

Weisskopf
A unit used to express the transition probability of nuclei from one state to another.

Wien's Constant

$$\lambda_m T = \text{constant} = 0{\cdot}28978 \text{ cm K}$$

(*See also* **Wien's Displacement Laws**.)

Zeeman Displacement
The displacement in wave-numbers for unit magnetic field given in SI by

$$\frac{e}{4\pi mc} = 46{\cdot}686 \text{ m}^{-1}\text{T}^{-1}.$$

(The formula is $e/4\pi mc^2$ if e is in e.s.u.) (*See also* **Zeeman Effect**.)

Table of organic compounds

In this table a list of the organic compounds mentioned in the text is given, along with their systematic names. These names have the advantage that they indicate the structure of the compound unequivocally for the benefit of those readers who are not *au fait* with organic chemistry. As a further aid the structural formulae are also given.

Name of compound	*Synonym*	*Structural formula*
Acetaldehyde	Ethanol	CH_3CHO
Acetic acid	Ethanoic acid	CH_3COOH
Acetoacetic ester	3-Oxo-butanoic acid	CH_3COCH_2COOH
Acetic Anhydride	Ethanoic anhydride	$(CH_3COO)_2CO$
Acetone	2-Propanone	CH_3COCH_3
Adipic acid	Hexandioic acid	$HOOC.(CH_2)_4.COOH$
Allyl alcohol	2-Propen-1-ol	$CH_2{:}CHCH_2OH$
o-Amino-benzaldehyde		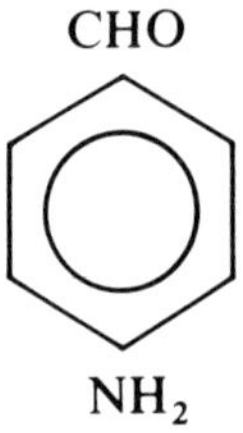 CHO NH_2
Aniline	Aminobenzene	$C_6H_5NH_2$
Anthraquinone	9, 10-Dioxy-anthracene	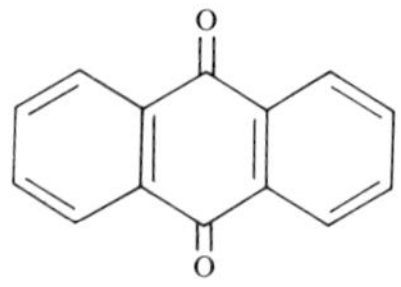 O O

Name of compound	*Synonym*	*Structural formula*
Anthraquinone β-sulphonate Na salt	9, 10-Dioxy-anthracene-2-sulphonate Na salt	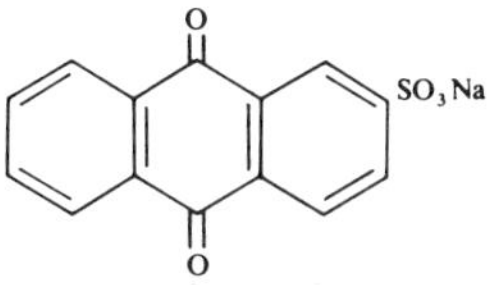 O SO_3Na O
Azoxybenzene		$C_6H_5.NO:N.C_6H_5$
Benzaldehyde		C_6H_5CHO
Benzene		C_6H_6

$C_6H_6 \equiv \quad \equiv$

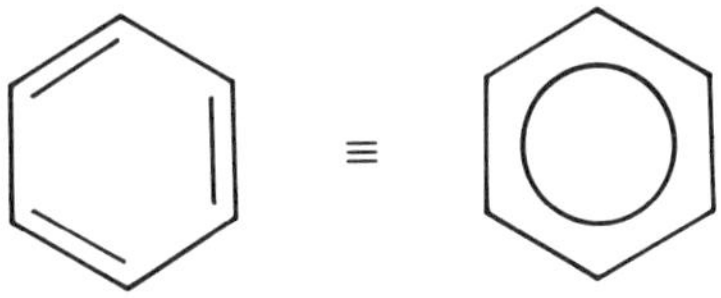

Benzoic acid		C_6H_5COOH
Benzotrichloride	ωωω-Trichloro-toluene	$C_6H_5CCl_3$
Benzoyl chloride		C_6H_5COCl
Benzyl alcohol	Phenyl carbinol	$C_6H_5CH_2OH$
Benzyl chloride	ω-Chlorotoluene	$C_6H_5CH_2Cl$
Benzyl cyanide	Phenylacetonitrile	$C_6H_5CH_2CN$
Butadiene	Divinyl	$CH_2{:}CHCH{:}CH_2$
n-Butane	Diethyl	$CH_3CH_2CH_2CH_3$
t-Butanol	2-Methylpropanol-2	$(CH_3)_3COH$
c-Butanol		$_2HC—CHOH$ $\vert \quad \vert$ $_2HC—CH_2$
c-Butylamine		$_2HC—CHNH_2$ $\vert \quad \vert$ $_2HC—CH_2$
t-Butoxide K salt		$(CH_3)_3CHOK$

Name of compound	*Synonym*	*Structural formula*
Carbylamine		C_2H_5NC
Chloroacetic acid		$CH_2ClCOOH$
Chloroacetone		CH_3COCH_2Cl
Cinchonic acid	4-Quinoline carboxylic acid	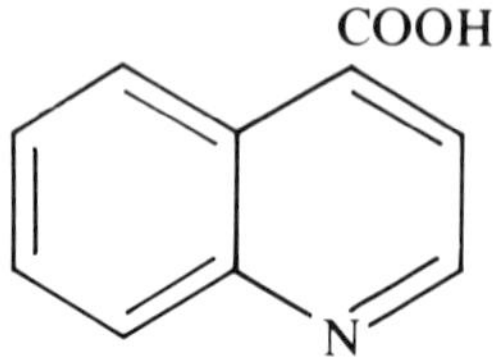 COOH N
Cinnamic acid	2-Phenyl acrylic acid	$C_6H_5CH{:}CHCOOH$
Crotonaldehyde	2-Butene-1-al	$CH_3CH{:}CHCHO$
Cyclopropyl-methylamine		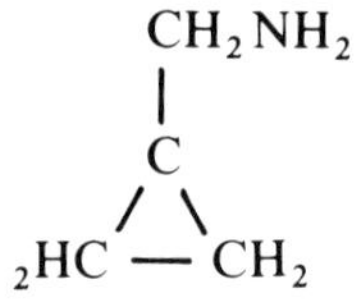 CH_2NH_2 C $_2HC$ — CH_2
Decalin	Decahydro-naphthalene	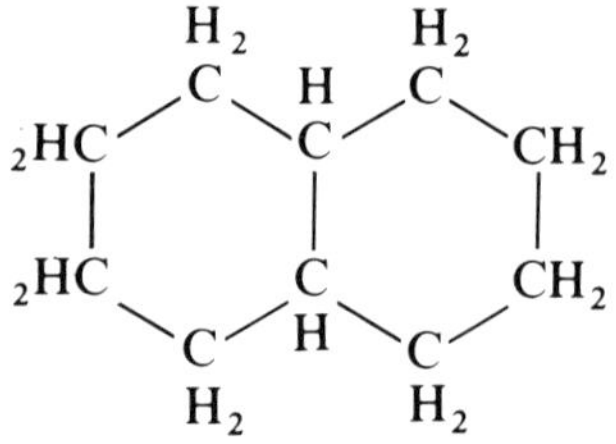 H_2 H_2 C H C $_2HC$ C CH_2 $_2HC$ C CH_2 C H C H_2 H_2 *cis* and *trans* forms
Diaminobenzoic acid	Four isomers 2, 3; 2, 4; 2, 5; and 3, 4-Diaminobenzoic acid	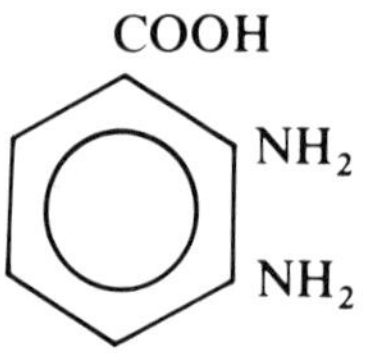 COOH NH_2 NH_2 2, 3 isomer
Diazomethane		CH_2N_2
Dibenzyl	1, 2-Diphenyl ethane	$C_6H_5.CH_2.CH_2.C_6H_5$

Name of compound	*Synonym*	*Structural formula*
Diethyl malonate	Malonic ester	$CH_2(COOC_2H_5)_2$
3, 4-Dihydro-isoquinoline		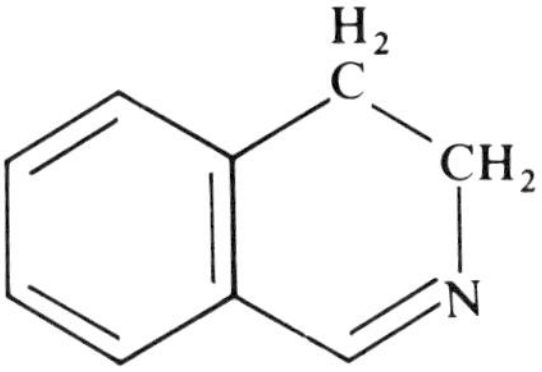 H_2 C, CH_2, N
Dinitrophenol	Four isomers (see diamino-benzoic acid)	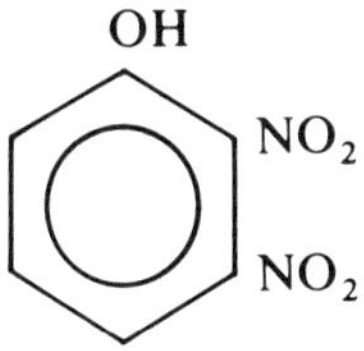 OH, NO_2, NO_2 2, 3 isomer
Dioxane		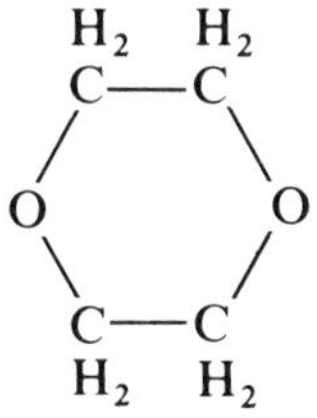 H_2 H_2 C—C O O C—C H_2 H_2
Diphenyl	Xenene	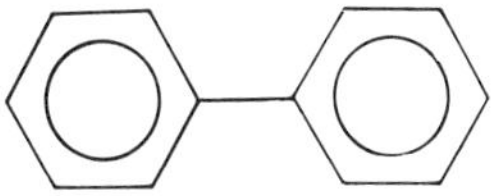
Diphenylamine		$C_6H_5NHC_6H_5$
Ethyl alcohol	Ethanol	C_2H_5OH
Ethoxide (Na salt)		C_2H_5ONa
Ethyl acetate		$CH_3COOC_2H_5$
Ethyl benzoate		$C_6H_5COOC_2H_5$
Ethyl cyanoacetate		$NCCH_2COOC_2H_5$
Ethyl formate		$HCOOC_2H_5$
Ethyl orthoformate	Aethon	$HC(OC_2H_5)_3$

Name of compound	*Synonym*	*Structural formula*
Ethyl oxalate	Diethyl oxalate	$(COOC_2H_5)_2$
Ethyl phenyl acetate		$C_6H_5CH_2COOC_2H_5$
Formaldehyde	Methanal, formalin	HCHO
Fructose		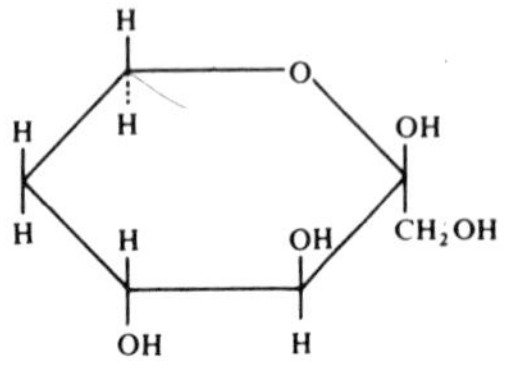
Fumaric acid	Butendioic acid	$(:CHCOOH)_2$
Glucose	Hexose	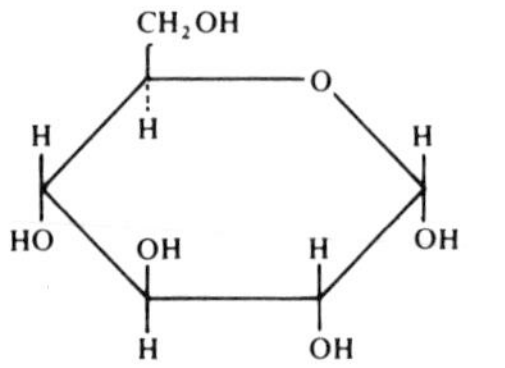
Glutaric acid	Pentandioic acid	$CH_2(CH_2COOH)_2$
Glycerol	Glycerine	$CHOH.(CH_2OH)_2$
Glycine	Aminoacetic acid	NH_2CH_2COOH
Hexamethylene tetramine	Methanamine	$(CH_2)_6N_4$
c-Hexane		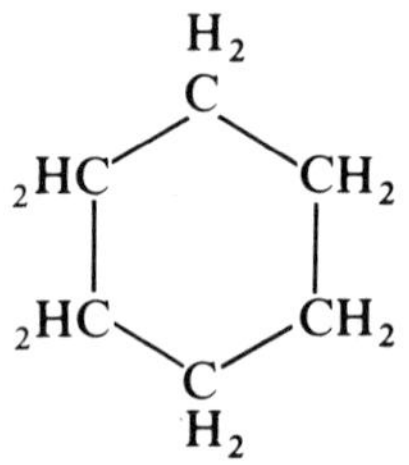

Name of compound	*Synonym*	*Structural formula*
p-Hydroxy-azobenzene		HO–C_6H_4–N = N–C_6H_5
o-Hydroxy-benzaldehyde	3-Hydroxy-benzaldehyde	CHO, OH (on benzene ring)
Indole	Benzopyrrole	(benzene ring fused to five-membered ring with NH)
Iodobenzene		C_6H_5I
Isatin		(benzene ring fused to ring with C = O, COH, N)
Isopropanol	Propanol-2	$(CH_3)_2CHOH$
Lactic acid	2-Hydroxy-propionic acid	$CH_3.CHOH.COOH$
Methane		CH_4
Malic acid	Hydroxysuccinic acid	H_2CCOOH / | / HOCHCOOH

Table of organic compounds

Name of compound	*Synonym*	*Structural formula*
Mannose		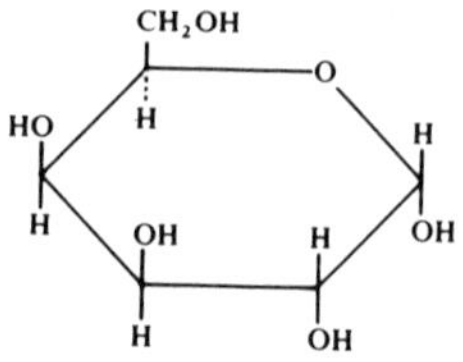
Methyl quinoline	Quinaldine	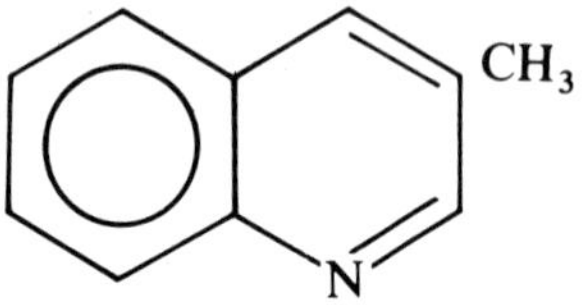
Naphthalene		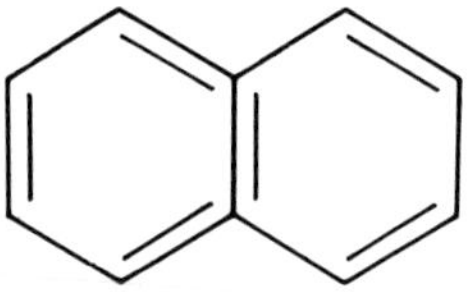
α-Naphthol		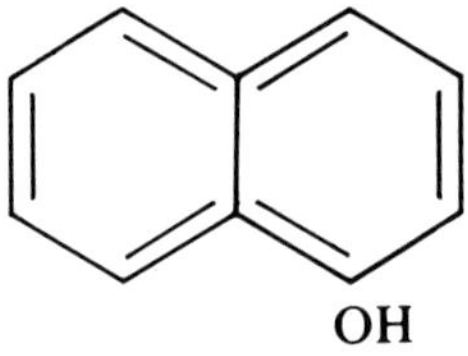
o-Nitro-benzaldehyde		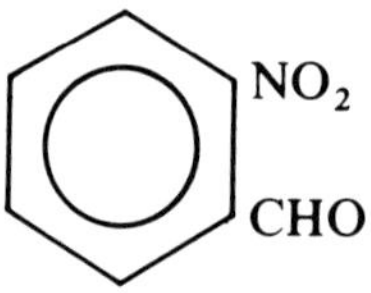
Nitrobenzene		$C_6H_5NO_2$
o-Nitrosophenol		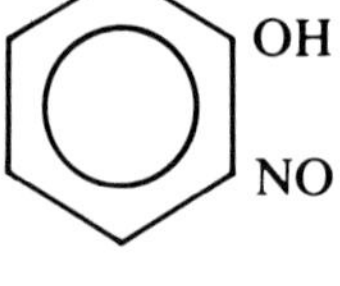

Name of compound	*Synonym*	*Structural formula*
p-Nitrosophenol		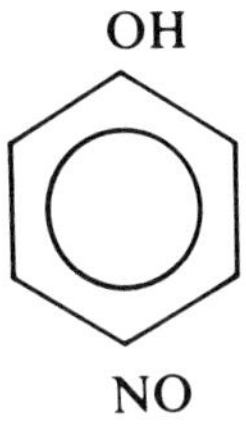 OH / benzene ring / NO
Oxalic acid	Ethandioic acid	$(COOH)_2$
Oxalacetic acid		$HOOCCH_2COCOOH$ two forms
Paraldehyde		$(CH_3CHO)_3$
Paraformaldehyde		$(HCHO)_nH_2O$
c-Pentadiene		 H / cyclopentadiene ring
Perbenzoic acid		C_6H_5COOOH
Perphthalic acid		$C_6H_4.COOH.COOOH$
Phenanthrene		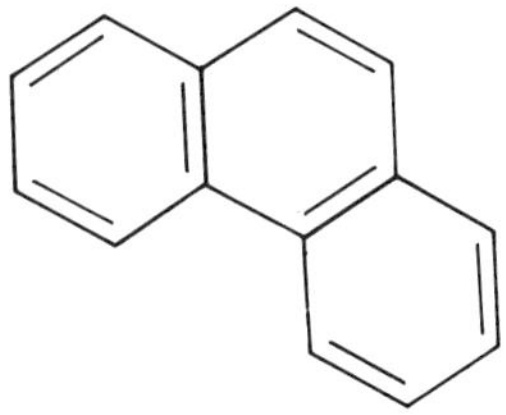 phenanthrene ring system
Phenanthrene-9-carboxylic acid	9-Phenanthroic acid	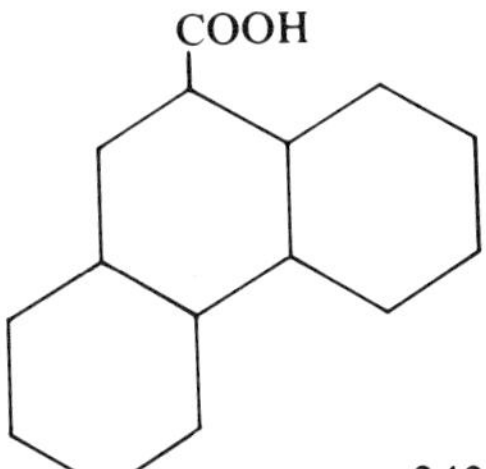 COOH on phenanthrene ring system

Name of compound	*Synonym*	*Structural formula*
Phenol		C_6H_5OH
Phenyl acetic acid		$C_6H_5CH_2COOH$
Phenyl aceturic acid		$C_6H_5CH_2CONHCH_2COOH$
Phenylene diamine	Diaminobenzene (1, 2; 1, 3; 1, 4)	$C_6H_4(NH_2)_2$
Phenylhydrazine		$C_6H_5NH{:}NH_2$
Phenyl malonic acid		$C_6H_5CH(COOH)_2$
α-Phenylo-nitro-cinammic acid		CH = CHCOOH NO_2
Phthalic acid	1, 2-Benzene dicarboxylic acid	$C_6H_4(COOH)_2$
Phthalimide	1, 3-Isoindole dione	O ‖ C NH C ‖ O
Picric acid	2, 4, 6-Trinitro-phenol	NO_2 $_2ON$ NO_2

Name of compound	*Synonym*	*Structural formula*
Pimelic acid	Heptandioic acid	$(CH_2)_5(COOH)_2$
Pinacol	2, 3-Butanediol-2, 3-dimethyl	$(CH_3)_2COHCOH(CH_3)_2$
Pinacolone	2-Butanone-3, 3-dimethyl	$(CH_3)_3CCOCH_3$
Propylene	Propene	$CH_3.CH:CH_2$
Pyridine	Azine	[six-membered ring with N]
Pyrrole	Azole	[five-membered ring with NH]
Quinoline		[fused bicyclic ring with N]
Resorcinol	1, 3-Dihydroxy benzene	$C_6H_4(OH)_2$
Rosaniline	(4-Amino-3-methylphenyl) -*bis* (4-amino-phenyl)-methanol	$(CH_3C_6H_3NH_2).C(OH)(C_6H_4NH_2)_2$
Succinic acid	Butandioic acid	$(CH_2COOH)_2$
Salicylic acid	2-Hydroxy-benzoic acid	COOH, OH [benzene ring]

Table of organic compounds

Name of compound	*Synonym*	*Structural formula*
Thiophen		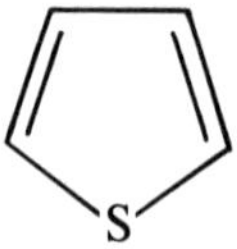
Toluene	Methylbenzene	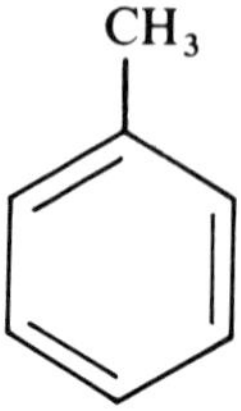
p-Toluene sulphonyl chloride		$C_6H_4(CH_3)SO_2Cl$
Trifluoroperoxy-acetic acid		CF_3COOOH
Triphenylamine		$(C_6H_5)_3N$
Xylene	1, 2; 1, 3; 1, 4-dimethylbenzene	$C_6H_4(CH_3)_2$